Andrew H Baker

Primary Lessons in Arithmetic

Andrew H Baker

Primary Lessons in Arithmetic

ISBN/EAN: 9783741129872

Manufactured in Europe, USA, Canada, Australia, Japa

Cover: Foto ©Thomas Meinert / pixelio.de

Manufactured and distributed by brebook publishing software
(www.brebook.com)

Andrew H Baker

Primary Lessons in Arithmetic

PRIMARY LESSONS

IN

ARITHMETIC.

BY

ANDREW H. BAKER, A.M., PH. D.,

AUTHOR OF "BAKER'S SERIES OF MATHEMATICS."

NEW YORK:

P. O'SHEA, PUBLISHER,

87 BARCLAY STREET.

1878.

SMITH & McDOUGAL, ELECTROTYPERS,
82 Beekman St., N. Y.

PREFACE.

THIS little book is intended for children beginning Arithmetic. It is better that a child should not engage in this study before the age of nine years, except in rare instances.

The first questions are such as children can readily answer without any previous study, and the progress is so gradual that they are prepared for each successive lesson as they advance.

Generally a young child does not prepare his lessons beforehand. The teacher must first lead the child carefully, without hurrying him faster than he can understand. The child soon learns to follow, and gradually to solve problems for himself. A few words of instruction will enable the child to understand and answer correctly all the questions in this book.

The child, however, should have a slate and pencil at his desk with which to amuse himself in making figures, or doing little sums at his own pleasure, not as a part of his tasks.

The lessons should not be too long, and sufficient time

should be allowed in their recital, that the child be not hurried, and the lessons should be repeated until they are indelibly impressed on the child's mind.

The Addition and Subtraction Table and the Multiplication and Division Table should be studied over and over again, till they can be recited without fault.

If the child hesitates or blunders in answering a question, it shows that he has not yet mastered the subject, and he should be made to repeat the lesson before taking another.

The Author cannot too highly recommend to the Teacher the use of the Blackboard described on the following page. Great facility in comprehending the combinations and divisions of numbers will be acquired by this method.

BLACKBOARD EXERCISE.

1   This page represents a blackboard with the num-   37
2   bers as high as 72 painted on its margins.   38
3   There is also a box containing slips which will cover   39
4   two, three, four, etc., as high as 12, and numbered   40
5   accordingly ; one of these the student will take in his   41
6   hand and apply it to the painted numbers to perform   42
7   addition or subtraction ; thus, begin at 1 and take a   43
8   slip marked 2, then 1 and 2 are 3, 3 and 2 are 5, 5 and   44
9   2 are 7, 7 and 2 are 9, etc., counting at least the left-   45
10   hand column ; then, to perform subtraction, begin at   46
11   the bottom of the 1st column; thus, 36 minus 2 equals   47
12   34, 34−2=32, 32−2=30, 30−2=28, etc., until the top   48
13   is reached ; then taking a slip marked 3, begin with   49
14   1 or 2, or first with 1 and then with 2, and return to   50
15   the top of the column as before, by subtraction ; let   51
16   this exercise be performed with all the slips, and as   52
17   the larger numbers are taken, continue the additions   53
18   to the bottom of the 2d column, and return as before.   54
19   For multiplication and division first make a chalk   55
20   mark after every two figures up to 24, and multi-   56
21   tiply ; thus, once 2 are 2, twice 2 are 4, 3 times 2 are   57
22   6, 4 times 2 are 8, etc. ; then the number of divisions   58
23   is 12 and each division has 2 numbers ; ∴ 12 is con-   59
24   tained twice in 24, or 2 is contained 12 times, 2 is   60
25   contained once in 2, in 4 twice, in 6 three times, in 8   61
26   four times, in 10 five times, in 12 six times, etc. When   62
27   the student is familiar with multiplication and division   63
28   by 2, let the numbers be separated into 3's, then 4's,   64
29   etc., and let each be continued for 12 divisions ; when   65
30   all the divisions have been performed according to the   66
31   steps, beginning with 2 and ending with 12, a multi-   67
32   plication and division table will be made.   68
33   REM.—In multiplication the product of any two fac-   69
34   tors is the same by making either the multiplicand and   70
35   the other the multiplier ; so also in division, the divisor   71
36   and the quotient may be substituted, as the dividend   72

is the product of the divisor and quotient.

REM.—The numbers, continued up to 144, should be painted on the sides of the board.

PRIMARY ARITHMETIC.

ADDITION AND SUBTRACTION.

1. I have two baskets with figs; the one has 10 and the other 5; if I count 5 from the larger number into the smaller, how many will there be in each basket?

2. Five and one are six, and one are seven, and one are eight, and one are nine, and one are ten. In the smaller basket there are now 10, and in the larger one only five remain; for, 1 from 10, 9 remain; 2 from 10, 8 remain; 3 from 10, 7 remain; 4 from 10, 6 remain; and 5 from 10, 5 remain.

Increasing the number in the one basket is ***Addition;*** and diminishing the number in the other basket is ***Subtraction.***

Count 30 and make the figures on the blackboard:

1,   2,   3,   4,   5,   6,   7,   8,   9; 10,

11, 12, 13, 14, 15, 16, 17, 18, 19, 20,

21, 22, 23, 24, 25, 26, 27, 28, 29, 30.

ADDITION AND SUBTRACTION TABLES.

LESSON I.

QUESTIONS FOR CLASS.

How many are 1 and 1?

How many are 1 and 2?

How many are 1 and 3 ?

How many are 1 and 4 ?

How many are 1 and 5 ?

How many are 1 and 6 ?

How many are 1 and 7 ?

How many are 1 and 8 ?

How many are 1 and 9 ?

How many are 2 and 1 ?

How many are 2 and 2 ?

How many are 2 and 3 ?

How many are 2 and 4 ?

How many are 2 and 5 ?

How many are 2 and 6 ?

How many are 2 and 7 ?

How many are 2 and 8 ?

How many are 2 and 9 ?

How many are 3 and 1 ?

How many are 3 and 2?

How many are 3 and 3 ?

How many are 3 and 4 ?

How many are 3 and 5 ?

How many are 3 and 6 ?

How many are 3 and 7 ?

How many are 3 and 8 ?

How many are 3 and 9 ?

How many are 4 and 1 ?

How many are 4 and 2 ?

How many are 4 and 3 ?

How many are 4 and 4 ?

How many are 4 and 5 ?

How many are 4 and 6 ?

How many are 4 and 7 ?

How many are 4 and 8 ?

How many are 4 and 9 ?

How many are 5 and 1 ?

How many are 5 and 2 ?

How many are 5 and 3 ?

How many are 5 and 4 ?

How many are 5 and 5 ?

How many are 5 and 6 ?

How many are 5 and 7 ?

How many are 5 and 8 ?

How many are 5 and 9 ?

LESSON II.

How many are 6 and 1 ? How many are 8 and 1 ?
How many are 6 and 2 ? How many are 8 and 2 ?
How many are 6 and 3 ? How many are 8 and 3 ?
How many are 6 and 4 ? How many are 8 and 4 ?
How many are 6 and 5 ? How many are 8 and 5 ?
How many are 6 and 6 ? How many are 8 and 6 ?
How many are 6 and 7 ? How many are 8 and 7 ?
How many are 6 and 8 ? How many are 8 and 8 ?
How many are 6 and 9 ? How many are 8 and 9 ?
How many are 7 and 1 ? How many are 9 and 1 ?
How many are 7 and 2 ? How many are 9 and 2 ?
How many are 7 and 3 ? How many are 9 and 3 ?
How many are 7 and 4 ? How many are 9 and 4 ?
How many are 7 and 5 ? How many are 9 and 5 ?
How many are 7 and 6 ? How many are 9 and 6 ?
How many are 7 and 7 ? How many are 9 and 7 ?
How many are 7 and 8 ? How many are 9 and 8 ?
How many are 7 and 9 ? How many are 9 and 9 ?

REM.—Let the student keep in mind that he is counting figs.

LESSON III.

How many remain ?

From 9 take 1. From 9 take 6. From 10 take 2.
From 9 take 2. From 9 take 7. From 10 take 3.
From 9 take 3. From 9 take 8 From 10 take 4.
From 9 take 4. From 9 take 9. From 10 take 5.
From 9 take 5. From 10 take 1. From 10 take 6.

From 10 take 7.	From 12 take 5.	From 14 take 3.
From 10 take 8.	From 12 take 6.	From 14 take 4.
From 10 take 9.	From 12 take 7.	From 14 take 5.
From 11 take 1.	From 12 take 8.	From 14 take 6.
From 11 take 2.	From 12 take 9.	From 14 take 7.
From 11 take 3.	From 13 take 1.	From 14 take 8.
From 11 take 4.	From 13 take 2.	From 14 take 9.
From 11 take 5.	From 13 take 3.	From 15 take 1.
From 11 take 6.	From 13 take 4.	From 15 take 2.
From 11 take 7.	From 13 take 5.	From 15 take 3.
From 11 take 8.	From 13 take 6.	From 15 take 4.
From 11 take 9.	From 13 take 7.	From 15 take 5.
From 12 take 1.	From 13 take 8.	From 15 take 6.
From 12 take 2.	From 13 take 9.	From 15 take 7.
From 12 take 3.	From 14 take 1.	From 15 take 8.
From 12 take 4.	From 14 take 2.	From 15 take 9.

LESSON IV.

From 16 take 1.	From 17 take 4.	From 18 take 7.
From 16 take 2.	From 17 take 5.	From 18 take 8.
From 16 take 3.	From 17 take 6.	From 18 take 9.
From 16 take 4.	From 17 take 7.	From 19 take 1.
From 16 take 5.	From 17 take 8.	From 16 take 2.
From 16 take 6.	From 17 take 9.	From 19 take 3.
From 16 take 7.	From 18 take 1.	From 19 take 4.
From 16 take 8.	From 18 take 2.	From 19 take 5.
From 16 take 9.	From 18 take 3.	From 19 take 6.
From 17 take 1.	From 18 take 4.	From 19 take 7.
From 17 take 2.	From 18 take 5.	From 19 take 8.
From 17 take 3.	From 18 take 6	From 19 take 9.

If you have five figs and give away 5, how many have you left?

If from any number you take away as many as the number itself, is there any left?

If you have only 5 pears, can you give away 6?

Can you take away from any number more than the number itself?

REM. 1.—In addition, we can unite any two numbers and make but one. Ex. 9 and 2 are 11, 9 and 3 are 12.

REM. 2.—In subtraction, we cannot take a larger number from a smaller one. Ex. 2 from 9 leaves 7, and 3 from 9 leaves 6.

REM. 3.—In the addition and subtraction of the same two numbers, the difference of the resulting numbers will be twice the smaller number. Ex. 6 and 2 are 8; 6 less 2 are 4; difference of results is 4, which is twice 2, the smaller number.

REM. 4.—The Subtraction Table should be read: Take 1 from 10, how many remain? 2 from 10? 3 from 10? 4 from 10? etc.

These tables should be recited repeatedly and distinctly, so as to be heard by the whole class; as beginners learn more by hearing others than by reciting themselves.

———◆◆◆———

LESSON V.

Count 50, and write the numbers on the blackboard.

How many are

10 and 1?	10 and 8?	11 and 5?	12 and 2?
10 and 2?	10 and 9?	11 and 6?	12 and 3?
10 and 3?	10 and 10?	11 and 7?	12 and 4?
10 and 4?	11 and 1?	11 and 8?	12 and 5?
10 and 5?	11 and 2?	11 and 9?	12 and 6?
10 and 6?	11 and 3?	11 and 10?	12 and 7?
10 and 7?	11 and 4?	12 and 1?	12 and 8?

12 and 9 ?	14 and 2 ?	15 and 5 ?	16 and 8 ?
12 and 10 ?	14 and 3 ?	15 and 6 ?	16 and 9 ?
13 and 1 ?	14 and 4 ?	15 and 7 ?	16 and 10 ?
13 and 2 ?	14 and 5 ?	15 and 8 ?	17 and 1 ?
13 and 3 ?	14 and 6 ?	15 and 9 ?	17 and 2 ?
13 and 4 ?	14 and 7 ?	15 and 10 ?	17 and 3 ?
13 and 5 ?	14 and 8 ?	16 and 1 ?	17 and 4 ?
13 and 6 ?	14 and 9 ?	16 and 2 ?	17 and 5 ?
13 and 7 ?	14 and 10 ?	16 and 3 ?	17 and 6 ?
13 and 8 ?	15 and 1 ?	16 and 4 ?	17 and 7 ?
13 and 9 ?	15 and 2 ?	16 and 5 ?	17 and 8 ?
13 and 10 ?	15 and 3 ?	16 and 6 ?	17 and 9 ?
14 and 1 ?	15 and 4 ?	16 and 7 ?	17 and 10 ?

LESSON VI.

In *Addition* we use the sign +, called *plus,* which indicates that the right-hand number is to be added to the left.

In *Subtraction* we use the sign —, called *minus,* which indicates that the right-hand number is to be subtracted from the left.

20 — 1 ? is read: 20 less 1, or 1 from 20 equals how many ?

20 — 1 ?	20 — 7 ?	21 — 3 ?	21 — 9 ?
20 — 2 ?	20 — 8 ?	21 — 4 ?	21 — 10 ?
20 — 3 ?	20 — 9 ?	21 — 5 ?	22 — 1 ?
20 — 4 ?	20 — 10 ?	21 — 6 ?	22 — 2 ?
20 — 5 ?	21 — 1 ?	21 — 7 ?	22 — 3 ?
20 — 6 ?	21 — 2 ?	21 — 8 ?	22 — 4 ?

22 — 5 ?	23 — 9 ?	25 — 3 ?	26 — 7 ?
22 — 6 ?	23 — 10 ?	25 — 4 ?	26 — 8 ?
22 — 7 ?	24 — 1 ?	25 — 5 ?	26 — 9 ?
22 — 8 ?	24 — 2 ?	25 — 6 ?	26 — 10 ?
22 — 9 ?	24 — 3 ?	25 — 7 ?	27 — 1 ?
22 — 10 ?	24 — 4 ?	25 — 8 ?	27 — 2 ?
23 — 1 ?	24 — 5 ?	25 — 9 ?	27 — 3 ?
23 — 2 ?	24 — 6 ?	25 — 10 ?	27 — 4 ?
23 — 3 ?	24 — 7 ?	26 — 1 ?	27 — 5 ?
23 — 4 ?	24 — 8 ?	26 — 2 ?	27 — 6 ?
23 — 5 ?	24 — 9 ?	26 — 3 ?	27 — 7 ?
23 — 6 ?	24 — 10 ?	26 — 4 ?	27 — 8 ?
23 — 7 ?	25 — 1 ?	26 — 5 ?	27 — 9 ?
23 — 8 ?	25 — 2 ?	26 — 6 ?	27 — 10 ?

LESSON VII.

18 and 1 ?	19 and 5 ?	20 and 9 ?	28 — 3 ?
18 and 2 ?	19 and 6 ?	20 and 10 ?	28 — 4 ?
18 and 3 ?	19 and 7 ?	21 and 1 ?	28 — 5 ?
18 and 4 ?	19 and 8 ?	21 and 2 ?	28 — 6 ?
18 and 5 ?	19 and 9 ?	21 and 3 ?	28 — 7 ?
18 and 6 ?	19 and 10 ?	21 and 4 ?	28 — 8 ?
18 and 7 ?	20 and 1 ?	21 and 5 ?	28 — 9 ?
18 and 8 ?	20 and 2 ?	21 and 6 ?	28 — 10 ?
18 and 9 ?	20 and 3 ?	21 and 7 ?	29 — 1 ?
18 and 10 ?	20 and 4 ?	21 and 8 ?	29 — 2 ?
19 and 1 ?	20 and 5 ?	21 and 9 ?	29 — 3 ?
19 and 2 ?	20 and 6 ?	21 and 10 ?	29 — 4 ?
19 and 3 ?	20 and 7 ?	28 — 1 ?	29 — 5 ?
19 and 4 ?	20 and 8 ?	28 — 2 ?	29 — 6 ?

29 — 7 ?	30 — 3 ?	30 — 9 ?	31 — 5 ?
29 — 8 ?	30 — 4 ?	30 — 10 ?	31 — 6 ?
29 — 9 ?	30 — 5 ?	31 — 1 ?	31 — 7 ?
29 — 10 ?	30 — 6 ?	31 — 2 ?	31 — 8 ?
30 — 1 ?	30 — 7 ?	31 — 3 ?	31 — 9 ?
30 — 2 ?	30 — 8 ?	31 — 4 ?	31 — 10 ?

After the students answer all the preceding questions readily and without fault, but not until then, they should be carefully cross-questioned until they make no failure.

PRACTICAL QUESTIONS.

1. Willie has seven cents, Andrew nine cents, and John twelve cents; how many cents have the three boys ?

2. Three boys bought an orange for 6 cents; a lemon for 5 cents, and a pound of sugar for 12 cents; they gave the confectioner a 25 cent piece; how much change will they get ?

3. A lady gave some peaches to each of three boys; to the one she gave 7, the next 8, and the third 9 peaches; how many did they all get ?

4. A boy bought an apple for 5 cents, a pear for 8 cents, an orange for 9 cents, and a lemon for 6 cents; how many cents did he spend ?

5. A boy bought a chest for 25 cents, and had a lock put to it which cost 12 cents; he then sold the chest for 40 cents; what was his profit ?

6. Three boys were counting their money; the first had 14 cents, the second 9 cents and the third 7 cents; they played in the grass for some time and when they

returned they had in all only 21 cents; how many cents did they lose?

7. A man bought a barrel of sugar for 15 dollars, a barrel of flour for 8 dollars, and a barrel of corn for 3 dollars; what was the amount of the bill?

8. A farmer sold 3 bushels of wheat for 5 dollars, a ton of hay for 12 dollars, a cord of wood for 4 dollars, and a lot of potatoes for 9 dollars; what was the amount of his sales?

9. A butcher sold 300 pounds of beef for 24 dollars, 100 pounds of pork for 6 dollars, mutton for 9 dollars, and veal for 4 dollars; what was the amount of his sales?

10. A boy bought a slate for 25 cents, a pencil for 3 cents, a bottle of ink for 8 cents, pens for 6 cents, and paper for 10 cents; how many cents did he spend?

LESSON VIII.

How many are

8 and 7 and 5?	9 and 6 less 5 ?	$9 + 8 + 7?$
9 and 5 and 4?	$8 + 7 - 6?$	$8 + 4 + 9?$
7 and 6 and 8?	$9 + 8 - 7?$	$6 + 7 - 3?$
9 and 7 and 5?	$7 + 9 - 5?$	$9 + 4 - 2?$
8 and 9 and 6?	$5 + 4 - 7?$	$12 + 6 + 4?$
7 and 5 and 4?	6 and 4 less 3?	$13 + 6 + 3?$
8 and 7 and 3?	$8 + 5 - 4?$	$11 + 9 + 5?$
6 and 9 and 7?	$6 + 3 - 2?$	$9 + 2 + 8?$
5 and 8 and 6?	$4 + 9 - 5?$	$7 + 3 + 4?$
4 and 9 and 7?	$3 + 7 - 4?$	$3 + 2 + 8?$

LESSON IX.

5 and 8 are how many?	15 — 8 are how many?
6 and 7 are how many?	12 — 9 are how many?
8 and 3 are how many?	12 — 7 are how many?
7 and 2 are how many?	10 — 6 are how many?
3 and 7 are how many?	14 — 8 are how many?
4 and 8 are how many?	16 — 9 are how many?
6 and 5 are how many?	12 — 5 are how many?
8 and 7 are how many?	13 — 8 are how many?
9 and 3 are how many?	13 — 7 are how many?
7 and 5 are how many?	11 — 3 are how many?
6 and 4 are how many?	9 — 2 are how many?
8 and 6 are how many?	10 — 7 are how many?
9 and 7 are how many?	12 — 8 are how many?
5 and 7 are how many?	11 — 5 are how many?
13 — 5 are how many?	15 — 7 are how many?
13 — 6 are how many?	12 — 3 are how many?
11 — 8 are how many?	12 — 5 are how many?
9 — 7 are how many?	10 — 4 are how many?
10 — 3 are how many?	14 — 6 are how many?
12 — 4 are how many?	16 — 7 are how many?
11 — 6 are how many?	12 — 7 are how many?

REM. 1.—If from the sum of two numbers either number be subtracted, the remainder will be the other number.

REM. 2.—In addition we have given two numbers, to find their sum.

REM. 3.—In subtraction we have given the sum of two numbers and one of the numbers, to find the other number.

REM. 4.—When in addition two numbers are united into one, this one is called the sum of the two numbers.

REM. 5.—In subtraction the larger given number corresponds

to the sum in addition, and is called the Minuend; the smaller given number is called the Subtrahend, and it corresponds to one of the given numbers in addition; the difference, which is called the remainder, corresponds to the other given number in addition.

LESSON X.

Count 100 and make the figures on the blackboard; carry the count to 125 and tell how it continues.

1. How many are 3 and 9? 13 and 9? 23 and 9? 33 and 9? 43 and 9? 53 and 9? 63 and 9? 73 and 9? 83 and 9? 93 and 9?

2. How many are 4 and 9? 14 and 9? 24 and 9? 34 and 9? 44 and 9? 54 and 9? 64 and 9? 74 and 9? 84 and 9? 94 and 9?

3. How many are 4 and 8? 14 and 8? 24 and 8? 34 and 8? 44 and 8? 54 and 8? 64 and 8? 74 and 8? 84 and 8? 94 and 8?

4. How many are 5 and 8? 15 and 8? 25 and 8? 35 and 8? 45 and 8? 55 and 8? 65 and 8? 75 and 8? 85 and 8? 95 and 8?

5. How many are 5 and 9? 15 and 9? 25 and 9? 35 and 9? 45 and 9? 55 and 9? 65 and 9? 75 and 9? 85 and 9? 95 and 9?

6. How many are 6 and 7? 16 and 7? 26 and 7? 36 and 7? 46 and 7? 56 and 7? 66 and 7? 76 and 7? 86 and 7? 96 and 7?

7. How many are 6 and 8? 16 and 8? 26 and 8? 36 and 8? 46 and 8? 56 and 8? 66 and 8? 76 and 8? 86 and 8? 96 and 8?

8. How many are 6 and 9 ?   16 and 9 ?   26 and 9 ?
36 and 9 ?   46 and 9 ?   56 and 9 ?   66 and 9 ?   76 and 9 ?
86 and 9 ?   96 and 9 ?

9. How many are 7 and 7 ?   17 and 7 ?   27 and 7 ?
37 and 7 ?   47 and 7 ?   57 and 7 ?   67 and 7 ?   77 and 7 ?
87 and 7 ?   97 and 7 ?

10. How many are 7 and 8 ?   17 and 8 ?   27 and 8 ?
37 and 8 ?   47 and 8 ?   57 and 8 ?   67 and 8 ?   77 and 8 ?
87 and 8 ?   97 and 8 ?

11. How many are 8 and 3 ?   18 and 3 ?   28 and 3 ?
38 and 3 ?   48 and 3 ?   58 and 3 ?   68 and 3 ?   78 and 3 ?
88 and 3 ?   98 and 3 ?

12. How many are 9 and 2 ?   19 and 2 ?   29 and 2 ?
39 and 2 ?   49 and 2 ?   59 and 2 ?   69 and 2 ?   79 and 2 ?
89 and 2 ?   99 and 2 ?

13. How many are 9 and 3 ?   19 and 3 ?   29 and 3 ?
39 and 3 ?   49 and 3 ?   59 and 3 ?   69 and 3 ?   79 and 3 ?
89 and 3 ?   99 and 3 ?

14. How many are 9 and 4 ?   19 and 4 ?   29 and 4 ?
39 and 4 ?   49 and 4 ?   59 and 4 ?   69 and 4 ?   79 and 4 ?
89 and 4 ?   99 and 4 ?

NUMERATION.—A number, as 1, 2, or 3, having but
one figure, or the right-hand figure of a number, holds
the units place, or is of the first order.   A number
having two figures, as 12, 25, or 36, the left-hand figure
holds the tens place, or is of the second order.   A number having three figures, as 123, 256, or 372, the left-
hand figure is hundreds, or is of the third order.   Read
all the numbers.

REM.—Whenever the units in the terms are alike, then will
the units of the sums be alike ; 9 and 4 are equal to 4 and 9.

LESSON XI.

ADDITION.				SUBTRACTION.			
2	12	22	32	10	20	30	40
8	8	8	8	8	8	8	8
10	20	30	40	2	12	22	32
42	52	62	72	50	60	70	80
8	8	8	8	8	8	8	8
50	60	70	80	42	52	62	72

REM. 1.—Whenever the sum of the units reaches ten, 1 must be put in the order of tens.

REM. 2.—When the sum of two numbers is made the minuend, and either of the numbers the subtrahend, the difference will be the other number.

REM. 3.—When the number of units of the subtrahend is greater than that of the minuend, one ten of the minuend must be added to its units, and from their sum the units of the subtrahend subtracted ; but the tens of the minuend must be diminished by one, or what amounts to the same, one ten added to the sub-trahend and then subtracted from the minuend.

REM. 4.—The same process must be continued, if there are higher orders in the numbers, as the same relation exists between each succeeding order ; thus,

> 10 units　make　1 ten.
> 10 tens　　"　　1 hundred.
> 10 hundreds "　1 thousand, etc.

REM. 5.—A few examples of larger numbers may be exhibited on the blackboard.

ADDITION.

2	12	22	32	42	52	62	72
9	9	9	9	9	9	9	9
11	21	31	41	51	61	71	81
3	13	23	33	43	53	63	73
9	9	9	9	9	9	9	9
12	22	32	42	52	62	72	82
4	14	24	34	44	54	64	74
9	9	9	9	9	9	9	9
13	23	33	43	53	63	73	83
5	15	25	35	45	55	65	75
9	9	9	9	9	9	9	9
14	24	34	44	54	64	74	84
6	16	26	36	46	56	66	76
9	9	9	9	9	9	9	9
15	25	35	45	55	65	75	85
7	17	27	37	47	57	67	77
9	9	9	9	9	9	9	9
16	26	36	46	56	66	76	86
8	18	28	38	48	58	68	78
9	9	9	9	9	9	9	9
17	27	37	47	57	67	77	87
9	19	29	39	49	59	69	79
9	9	9	9	9	9	9	9
18	28	38	48	58	68	78	88

SUBTRACTION.

11	21	31	41	51	61	71	81
9	9	9	9	9	9	9	9
2	12	22	32	42	52	62	72

12	22	32	42	52	62	72	82
9	9	9	9	9	9	9	9
3	13	23	33	43	53	63	73

13	23	33	43	53	63	73	83
9	9	9	9	9	9	9	9
4	14	24	34	44	54	64	74

14	24	34	44	54	64	74	84
9	9	9	9	9	9	9	9
5	15	25	35	45	55	65	75

15	25	35	45	55	65	75	85
9	9	9	9	9	9	9	9
6	16	26	36	46	56	66	76

16	26	36	46	56	66	76	86
9	9	9	9	9	9	9	9
7	17	27	37	47	57	67	77

17	27	37	47	57	67	77	87
9	9	9	9	9	9	9	9
8	18	28	38	48	58	68	78

18	28	38	48	58	68	78	88
9	9	9	9	9	9	9	9
9	19	29	39	49	59	69	7ᶜ

ADDITION AND SUBTRACTION TABLE.

0	1	2	3	4	5	6	7	8	9
1	2	3	4	5	6	7	8	9	10
2	3	4	5	6	7	8	9	10	11
3	4	5	6	7	8	9	10	11	12
4	5	6	7	8	9	10	11	12	13
5	6	7	8	9	10	11	12	13	14
6	7	8	9	10	11	12	13	14	15
7	8	9	10	11	12	13	14	15	16
8	9	10	11	12	13	14	15	16	17
9	10	11	12	13	14	15	16	17	18
10	11	12	13	14	15	16	17	18	19
11	12	13	14	15	16	17	18	19	20
12	13	14	15	16	17	18	19	20	21
13	14	15	16	17	18	19	20	21	22
14	15	16	17	18	19	20	21	22	23
15	16	17	18	19	20	21	22	23	24
16	17	18	19	20	21	22	23	24	25
17	18	19	20	21	22	23	24	25	26
18	19	20	21	22	23	24	25	26	27
19	20	21	22	23	24	25	26	27	28
20	21	22	23	24	25	26	27	28	29
21	22	23	24	25	26	27	28	29	30

The square made by the ten Arabic characters forms an Addition and a Subtraction Table.

Beginning with the first line, thus, Zero and zero are zero; zero and 1 are 1; zero and 2 are 2; zero and 3 are 3; zero and 4 are 4; zero and 5 are 5; zero and 6 are 6, etc.

The second line, one and zero are one; 1 and 1 are 2; 1 and 2 are 3 ; 1 and 3 are 4; 1 and 4 are 5, etc.

2 and zero are 2; 2 and 1 are 3; 2 and 2 are 4; 2 and 3 are 5 ; 2 and 4 are 6, etc.

3 and 0 are 3 ; 3 and 1 are 4 ; 3 and 2 are 5; 3 and 3 are 6; 3 and 4 are 7, etc.

Continue this, taking the first figure of the 1st column and adding it to each successive figure in the first line ; the adding of zero is only nominal, as it makes no increase.

It also becomes a subtraction table, the figures of the first column being the subtrahend, and those of the first line the remainders.

Take zero from 1 and 1 remains; 0 from 2, 2 remain, etc. It may be thus expressed: $1 - 0 = 1$; $2 - 0 = 2$; $3 - 0 = 3$; $4 - 0 = 4$; $5 - 0 = 5$; which is read, 1 minus zero equals 1, etc.

Second line: take 1 from 2, 1 remains; 1 from 3, 2 remain; or, $2 - 1 = 1$; $3 - 1 = 2$; $4 - 1 = 3$; $5 - 1 = 4$, etc.

Third line: $3 - 2 = 1$; $4 - 2 = 2$; $5 - 2 = 3$; $6 - 2 = 4$; $7 - 2 = 5$, etc.

In addition, we add two numbers at a time, never more, and in the first square we have the addition of every two units that can come together ; so also in subtraction.

In the second square, the units correspond with the first square, and have an additional ten.

In the third square, the units again are repeated, and another additional ten.

As a column of tens, hundreds, and every higher or lower order is added and subtracted in the same way, the accompanying table develops every principle of addition and subtraction.

MULTIPLICATION AND DIVISION.

LESSON XII.

1. When apples are worth *1 cent apiece,* what will 1 apple cost? 2 apples? 3 apples? 4 apples? 5 apples? 6 apples? 7 apples? 8 apples? 9 apples? 10 apples? 11 apples? 12 apples?

2. At *2 cents apiece,* what will 1 apple cost? 2 apples? 3 apples? 4 apples? 5 apples? 6 apples? 7 apples? 8 apples? 9 apples? 10 apples? 11 apples? 12 apples?

3. At *3 cents apiece,* what will 1 apple cost? 2 apples? 3 apples? 4 apples? 5 apples? 6 apples? 7 apples? 8 apples? 9 apples? 10 apples? 11 apples? 12 apples?

4. At *4 cents apiece,* what will 1 apple cost? 2 apples? 3 apples? 4 apples? 5 apples? 6 apples? 7 apples? 8 apples? 9 apples? 10 apples? 11 apples? 12 apples?

5. At *5 cents apiece,* what will 1 orange cost? 2 oranges? 3 oranges? 4 oranges? 5 oranges? 6 oranges? 7 oranges? 8 oranges? 9 oranges? 10 oranges? 11 oranges? 12 oranges?

6. At *6 cents apiece,* what will 1 orange cost? 2 oranges? 3 oranges? 4 oranges? 5 oranges? 6 oranges? 7 oranges? 8 oranges? 9 oranges? 10 oranges? 11 oranges? 12 oranges?

LESSON XIII.

1. What will *2 yards of tape* cost at 2 cents a yard ? at 3 cents a yard? at 4 cents a yard ?

2. What will *3 yards of tape* cost at 2 cents a yard ? at 3 cents a yard? at 4 cents a yard ?

3. What will *4 yards of tape* cost at 2 cents a yard? at 3 cents a yard? at 4 cents a yard ?

4. What will *5 yards of tape* cost at 2 cents a yard ? at 3 cents a yard? at 4 cents a yard?

5. What will *6 yards of tape* cost at 2 cents a yard ? at 3 cents a yard ? at 4 cents a yard ?

6. What will *2 yards of tape* cost at 5 cents a yard? at 6 cents a yard ? at 7 cents a yard ?

7. What will *3 yards of tape* cost at 5 cents a yard? at 6 cents a yard? at 7 cents a yard ?

8. What will *4 yards of tape* cost at 5 cents a yard? at 6 cents a yard ? at 7 cents a yard ?

9. What will *5 yards of tape* cost at 5 cents a yard? at 6 cents a yard? at 7 cents a yard?

10. What will *6 yards of tape* cost at 5 cents a yard? at 6 cents a yard? at 7 cents a yard?

11. What will *2 yards of tape* cost at 8 cents a yard? at 9 cents a yard? at 10 cents a yard?

12. What will *3 yards of tape* cost at 8 cents a yard ? at 9 cents a yard ? at 10 cents a yard ?

13. What will *4 yards of tape* cost at 8 cents a yard ? at 9 cents a yard? at 10 cents a yard ?

14. What will *5 yards of tape* cost at 8 cents a yard? at 9 cents a yard? at 10 cents a yard ?

15. What will *6 yards of tape* cost at 8 cents a yard ? at 9 cents a yard ? at 10 cents a yard ?

2

16. What will *7 yards of tape* cost at 8 cents a yard? at 9 cents a yard? at 10 cents a yard?

17. What will *8 yards of tape* cost at 8 cents a yard? at 9 cents a yard? at 10 cents a yard?

18. What will *9 yards of tape* cost at 8 cents a yard? at 9 cents a yard? at 10 cents a yard?

LESSON XIV.

1. At *1 cent apiece,* how many apples can you buy for 1 cent? for 2 cents? for 3 cents? for 4 cents? for 5 cents? for 6 cents?

2. At *2 cents apiece,* how many can you buy for 2 cents? for 4 cents? for 6 cents? for 8 cents? for 10 cents? for 12 cents?

3. At *3 cents apiece,* how many can you buy for 3 cents? for 6 cents? for 9 cents? for 12 cents? for 15 cents? for 18 cents?

4. At *4 cents apiece,* how many can you buy for 4 cents? for 8 cents? for 12 cents? for 16 cents? for 20 cents? for 24 cents?

5. At *5 cents apiece,* how many can you buy for 5 cents? for 10 cents? for 15 cents? for 20 cents? for 25 cents? for 30 cents?

6. At *6 cents apiece,* how many can you buy for 6 cents? for 12 cents? for 18 cents? for 24 cents? for 30 cents? for 36 cents?

7. At *2 cents apiece,* how many can you buy for 14 cents? for 16 cents? for 18 cents? for 20 cents? for 22 cents? for 24 cents?

8. At *3 cents apiece,* how many can you buy for 21 cents? for 24 cents? for 27 cents? for 30 cents? for 33 cents? for 36 cents?

9. At *4 cents apiece,* how many can you buy for 28 cents? for 32 cents? for 36 cents? for 40 cents? for 44 cents? for 48 cents?

10. At *5 cents apiece,* how many can you buy for 35 cents? for 40 cents? for 45 cents? for 50 cents? for 55 cents? for 70 cents?

11. At *6 cents apiece,* how many can you buy for 42 cents? for 48 cents? for 54 cents? for 60 cents? for 66 cents? for 72 cents?

———•◦•———

LESSON XV.

1. What would be the cost of 2 oranges at *7 cents apiece?* of 3 oranges? of 4 oranges? of 5 oranges? of 6 oranges? of 7 oranges? of 8 oranges? of 9 oranges? of 10 oranges? of 11 oranges? of 12 oranges?

2. Of 2 oranges at *8 cents apiece?* of 3 oranges? of 4 oranges? of 5 oranges? of 6 oranges? of 7 oranges? of 8 oranges? of 9 oranges? of 10 oranges? of 11 oranges? of 12 oranges?

3. Of 2 oranges at *9 cents apiece?* of 3 oranges? of 4 oranges? of 5 oranges? of 6 oranges? of 7 oranges? of 8 oranges? of 9 oranges? of 10 oranges? of 11 oranges? of 12 oranges?

4. Of 2 oranges at *10 cents apiece?* of 3 oranges? of 4 oranges? of 5 oranges? of 6 oranges? of 7 oranges? of 8 oranges? of 9 oranges? of 10 oranges? of 11 oranges? of 12 oranges?

5. What would be the cost of 2 oranges at *11 cents apiece?* of 3 oranges? of 4 oranges? of 5 oranges? of 6 oranges? of 7 oranges? of 8 oranges? of 9 oranges? of 10 oranges? of 11 oranges? of 12 oranges?

6. Of 2 oranges at *12 cents apiece?* of 3 oranges? of 4 oranges? of 5 oranges? of 6 oranges? of 7 oranges? of 8 oranges? of 9 oranges? of 10 oranges? of 11 oranges? of 12 oranges?

———◆◆◆———

LESSON XVI.

1. At *7 cents apiece,* how many melons can I buy for 14 cents? for 21 cents? for 28 cents? for 35 cents? for 42 cents? for 49 cents? for 56 cents? for 63 cents? for 70 cents? for 77 cents? for 84 cents?

2. At *8 cents apiece,* how many can I buy for 16 cents? for 24 cents? for 32 cents? for 40 cents? for 48 cents? for 56 cents? for 64 cents? for 72 cents? for 80 cents? for 88 cents? for 96 cents?

3. At *9 cents apiece,* how many can I buy for 18 cents? for 27 cents? for 36 cents? for 45 cents? for 54 cents? for 63 cents? for 72 cents? for 81 cents? for 90 cents? for 99 cents? for 108 cents?

4. At *10 cents apiece,* how many can I buy for 20 cents? for 30 cents? for 40 cents? for 50 cents? for 60 cents? for 70 cents? for 80 cents? for 90 cents? for 100 cents? for 110 cents? for 120 cents?

5. At *11 cents apiece,* how many can I buy for 22 cents? for 33 cents? for 44 cents? for 55 cents? for 66 cents? for 77 cents? for 88 cents? for 99 cents? for 110 cents? for 121 cents? for 132 cents?

6. At *12 cents apiece,* how many can I buy for 24 cents? for 36 cents? for 48 cents? for 60 cents? for 72 cents? for 84 cents? for 96 cents? for 108 cents? for 120 cents? for 132 cents? for 144 cents?

LESSON XVII.

MULTIPLICATION AND DIVISION TABLE.

1	2	3	4	5	6	7	8	9	10	11	12
2	4	6	8	10	12	14	16	18	20	22	24
3	6	9	12	15	18	21	24	27	30	33	36
4	8	12	16	20	24	28	32	36	40	44	48
5	10	15	20	25	30	35	40	45	50	55	60
6	12	18	24	30	36	42	48	54	60	66	72
7	14	21	28	35	42	49	56	63	70	77	84
8	16	24	32	40	48	56	64	72	80	88	96
9	18	27	36	45	54	63	72	81	90	99	108
10	20	30	40	50	60	70	80	90	100	110	120
11	22	33	44	55	66	77	88	99	110	121	132
12	24	36	48	60	72	84	96	108	120	132	144

As a Multiplication Table: Begin with the first line; thus, Once 1 is 1; once 2 are 2; once 3 are 3, etc. Second line, Once 2 are 2; twice 2 are 4; 3 times 2 are 6; 4 times 2 are 8, etc, Third line, Once 3 are 3; twice 3 are 6; 3 times 3 are nine; 4 times 3 are 12, etc. Recite the other lines in the same manner.

As a Division Table: Begin with the first line; thus, 1 is contained in 1, once; in 2, twice; in 3, three times; in 4, four times, etc. Second line, 2 into 2 = 1; 2 into 4 = 2; 2 into 6 = 3; 2 into 8 = 4, etc. Third line, 3 into 3 = 1; 3 into 6 = 2, etc.

REM.—As a Multiplication Table, it may also be read by the column, by which the factors are alternate, without changing the product. Any number is multiplied by 10 by adding a zero to it. As a Division Table, the first column has all the divisors, the first line all the quotients, and every number in each line is a dividend, which is always in the same line and the same column with the quotient and divisor. Any number having a zero in the units place is divided by 10 by removing the zero.

---***---

LESSON XVIII.

1. In 1 pint there are *4 gills.* In 2 pints, how many gills? in 3 pints? in 4 pints? in 8 pints?

2. How many *pints* in 4 gills? in 8 gills? in 12 gills? in 16 gills? in 32 gills?

3. In 1 quart there are *2 pints.* In 2 quarts, how many pints? in 3 quarts? in 4 quarts? in 10 quarts? in 12 quarts?

4. In 2 pints, how many *quarts?* in 4 pints? in 6 pints? in 8 pints? in 20 pints? in 24 pints?

5. There are *4 quarts* in 1 gallon. How many quarts in 5 gallons? in 8 gallons? in 12 gallons? in 16 gallons?

6. How many *gallons* in 4 quarts? in 20 quarts? in 32 quarts? in 48 quarts? in 64 quarts?

LESSON XIX.

1. There are 10 mills in 1 cent. How many mills in 3 cents ? in 4 cents ? in 5 cents ? in 6 cents ?

2. There are 10 cents in 1 dime. How many cents in 2 dimes ? in 3 dimes ? in 4 dimes ? in 5 dimes ?

3. There are 10 dimes in 1 dollar. How many dimes in 3 dollars ? in 5 dollars ? in 7 dollars ?

4. There are 100 cents in 1 dollar. How many cents in 2 dollars ? in 3 dollars ? in 5 dollars ?

5. How many cents in 10 mills ? How many cents in 20 mills ? How many in 50 mills ?

6. How many dimes in 10 cents ? in 20 cents ? in 30 cents ? in 40 cents ?

7. How many dollars in 10 dimes ? in 20 dimes ? in 30 dimes ? in 40 dimes ?

8. How many dollars in 100 cents ? in 200 cents ? in 300 cents ? in 500 cents ?

PRACTICAL QUESTIONS.

1. If 1 gill of beer cost 2 cents, what will a pint cost ? A quart ? A gallon ?

2. If 1 gill of wine cost 4 cents, what will a pint cost ? A quart ? A gallon ?

3. If two men can do a piece of work in 1 day, how long will it take 1 man to do it ?

Ans. It will take 1 man 2 days.

4. If one man can do a piece of work in 3 days. how long will it take 3 men to do it ? *Ans.* 1 day.

REM.—It is evident that it will take one man twice as long as 2 men ; and it is also evident that 3 men will do the same work in one-third the time in which 1 man will do it.

5. If 3 men can do a piece of work in 3 days, how long will it take 1 man to do it ?

6. If 1 man can do a piece of work in 9 days, how long will it take 3 men to do it ?

7. If 4 men can do a piece of work in 3 days, how long will it take 6 men to do it ?

LESSON XX.

1. What cost 2 yards of cloth at 5 dollars a yard ?
2. At $5 a yard, how many yards can you buy for $10 ?
3. What cost 3 yards of cloth at 6 dollars a yard ?
4. At $6 a yard, how many yards can you buy for $18 ?
5. What cost 4 yards of cloth at 7 dollars a yard ?
6. At $7 a yard, how many yards can you buy for $28 ?
7. What cost 5 yards of cloth at 8 dollars a yard ?
8. At $8 a yard, how many yards can you buy for $40 ?
9. What cost 6 yards of cloth at 9 dollars a yard ?
10. At $9 a yard, how many yards can you buy for $54 ?
11. What cost 2 yards of cloth at 4 dollars a yard ?
12. At $4 a yard, how many yards can you buy for $8 ?
13. What cost 3 yards of cloth at 5 dollars a yard ?
14. At $5 a yard, how many yards can you buy for $15 ?
15. What cost 4 yards of cloth at 6 dollars a yard ?
16. At $6 a yard, how many yards can you buy for $24 ?
17. What cost 5 yards of cloth at 7 dollars a yard ?
18. At $7 a yard, how many yards can you buy for $35 ?

19. What cost 6 yards of cloth at 8 dollars a yard ?
20. At $8 a yard, how many yards can you buy for $48 ?
21. What cost 2 yards of cloth at 3 dollars a yard ?
22. At $3 a yard, how many yards can you buy for $6 ?
23. What cost 3 yards of cloth at 4 dollars a yard ?
24. At $4 a yard, how many yards can you buy for $12 ?
25. What cost 4 yards of cloth at 5 dollars a yard ?
26. At $5 a yard, how many yards can you buy for $20 ?
27. What cost 5 yards of cloth at 6 dollars a yard ?
28. At $6 a yard, how many yards can you buy for $30 ?
29. What cost 6 yards of cloth at 7 dollars a yard ?
30. At $7 a yard, how many yards can you buy for $42 ?
31. What cost 2 yards of cloth at 2 dollars a yard ?
32. At $2 a yard, how many yards can you buy for $4 ?
33. What cost 3 yards of cloth at 3 dollars a yard ?
34. At $3 a yard, how many yards can you buy for $9 ?
35. What cost 4 yards of cloth at 4 dollars a yard ?
36. At $4 a yard, how many yards can you buy for $16 ?
37. What cost 5 yards of cloth at 5 dollars a yard ?
38. At $5 a yard, how many yards can you buy for $25 ?
39. What cost 6 yards of cloth at 6 dollars a yard ?
40. At $6 a yard, how many yards can you buy for $36 ?
41. What cost 2 yards of cloth at 6 dollars a yard ?
42. At $6 a yard, how many yards can you buy for $12 ?
43. What cost 3 yards of cloth at 7 dollars a yard ?
44. At $7 a yard, how many yards can you buy for $21 ?
45. What cost 4 yards of cloth at 8 dollars a yard ?
46. At $8 a yard, how many yards can you buy for $32 ?
47. What cost 5 yards of cloth at 9 dollars a yard ?
48. At $9 a yard, how many yards can you buy for $45 ?
49. What cost 6 yards of cloth at 10 dollars a yard ?
50. At $10 a yard, how many yards can you buy for $60 ?

LESSON XXI.

1. Two men start from the same place at the same time and travel in the same direction; the one goes 5 miles an hour and the other 3; how far apart will they be in 1 hour? in 2 hours? in 3 hours? in 4 hours? in 5 hours? in 6 hours? in 7 hours? in 8 hours? in 9 hours? in 10 hours? in 11 hours? in 12 hours?

2. Two men start from the same place at the same time and travel in opposite directions; the one goes 3 miles an hour and the other 2; how far apart will they be in 1 hour? in 2 hours? in 3 hours? in 4 hours? in 5 hours? in 6 hours? in 7 hours? in 8 hours? in 9 hours? in 10 hours? in 11 hours? in 12 hours?

3. A cistern has two pipes; by the one 7 gallons run into it in an hour, and by the other 3 gallons run out; how many gallons will be in the cistern in 1 hour? in 2 hours? in 3 hours? in 4 hours? in 5 hours? in 6 hours? in 7 hours? in 8 hours? in 9 hours? in 10 hours? in 11 hours? in 12 hours?

4. If 6 men can do a piece of work in 4 days, how long will it take 1 man to do it? 2 men? 3 men? 4 men? 8 men? 12 men?

5. If 4 horses eat 6 bushels of oats in 3 days, how long would they last 1 horse? 2 horses? 3 horses? 6 horses?

6. If 12 men can do a piece of work in 6 days, how long will it take one man to do it? 2 men? 3 men? 4 men? 6 men? 8 men? 9 men?

7. If 1 man can do a piece of work in 72 days, how long will it take 6 men to do it?

8. If 5 men can do a piece of work in 12 days, how long will it take 3 men to do it? 4 men? 6 men?

FRACTIONS.

LESSON XXII.

DEF. 1.—If any thing or number is divided into equal parts, these parts are called *Fractions* of the whole, and all the parts constitute the whole.

DEF. 2.—When any thing or number is divided into two equal parts, one of the parts is called *one-half* of the thing or number, and the two parts constitute the whole.

If an apple or an orange is divided into two equal parts, one of the parts is one-half.

If an apple is worth 2 cents, what is one-half of it worth?   *Ans.* 1 cent.

Why? Because, if 2 cents is divided into two equal parts, one of the parts is 1 cent.

DEF. 3.—When any thing or number is divided into 3 equal parts, one of the parts is called *one-third ;* two of the parts, *two-thirds ;* and the three parts constitute the whole.

DEF. 4.—When any thing or number is divided into four equal parts, one of the parts is called *one-fourth,* two are *two-fourths,* three are *three-fourths,* and the four parts constitute the whole.

DEF. 5.—*Fifths, sixths, sevenths, eighths,* etc., are formed in a similar way, and they are written as follows :

One-half	is written $\frac{1}{2}$.	One-fifth	is written $\frac{1}{5}$.
One-third	is written $\frac{1}{3}$.	Two-fifths	is written $\frac{2}{5}$.
Two-thirds	is written $\frac{2}{3}$.	Three-fifths	is written $\frac{3}{5}$.
One-fourth	is written $\frac{1}{4}$.	Four-fifths	is written $\frac{4}{5}$.
Two-fourths	is written $\frac{2}{4}$.	One-sixth	is written $\frac{1}{6}$.
Three-fourths	is written $\frac{3}{4}$.	Two-sixths	is written $\frac{2}{6}$.
	etc.		etc.

The lower term, called the **Denominator,** indicates the number of parts into which a thing or number is divided; and the upper term, called the **Numerator,** indicates the number of parts taken, or the number divided.

REM. 1.—In any case of division, the dividend is the numerator, and the divisor the denominator of a fraction. Thus, to divide 15 by 7; if placed in a fractional form, it becomes $\frac{15}{7}$, which fraction is the quotient.

REM. 2.—When the dividend is greater than the divisor, the quotient is an improper fraction; but when the dividend is less than the divisor, the quotient is a proper fraction.

LESSON XXIII.

If one-half of an apple is divided into two equal parts, it makes two fourths; if it is divided into 3 equal parts, it makes $\frac{3}{6}$; for, in the first case, if the whole apple be divided in that way, it would make four parts, and in the second, six parts, etc.; therefore,

$$\frac{1}{2} = \frac{2}{4} = \frac{3}{6} = \frac{4}{8} = \frac{5}{10} = \frac{6}{12} = \frac{7}{14}, \text{ etc.}$$
$$\frac{1}{3} = \frac{2}{6} = \frac{3}{9} = \frac{4}{12} = \frac{5}{15} = \frac{6}{18}, \text{ etc.}$$

REM.—If both terms of a fraction are either multiplied or divided by the same number, the value of the fraction is not changed; for, by dividing each term of the resulting fraction, the original fraction is restored.

1. What is one-half of two? *Ans.* 1.
2. One is what part of two? *Ans.* $\frac{1}{2}$.
3. What is one-third of three? *Ans.* 1.
4. One is what part of three? *Ans.* $\frac{1}{3}$.
5. What is two-thirds of three? *Ans.* 2.
6. Two is what part of three? *Ans.* $\frac{2}{3}$.
7. What is one-fourth of four? *Ans.* 1.
8. One is what part of four? *Ans.* $\frac{1}{4}$.
9. What is two-fourths of four? *Ans.* 2.
10. Two is what part of four? *Ans.* $\frac{2}{4}$.
11. What is three-fourths of four? *Ans.* 3.
12. Three is what part of four? *Ans.* $\frac{3}{4}$.
13. What is one-fifth of five? *Ans.* 1.
14. One is what part of five? *Ans.* $\frac{1}{5}$.
15. What is two-fifths of five? *Ans.* 2.
16. Two is what part of five? *Ans.* $\frac{2}{5}$.
17. What is three-fifths of five? *Ans.* 3.
18. Three is what part of five? *Ans.* $\frac{3}{5}$.
19. What is four-fifths of five? *Ans.* 4.
20. Four is what part of five? *Ans.* $\frac{4}{5}$.
21. $\frac{5}{5}$ are what? *Ans.* The whole.
22. If you can buy a pear for 2 cents, how many can you buy for 1 cent? *Ans.* $\frac{1}{2}$.
23. If you can buy a pear for 2 cents, how many can you buy for 3 cents? *Ans.* $\frac{3}{2} = 1\frac{1}{2}$.

Rem.—For 1 cent you can buy $\frac{1}{2}$ a pear, and for 3 cents, 3 times as much, that is, $\frac{3}{2}$, or one and one-half.

24. How many can you buy for 4 cents? for 5 cents? 6 cents? 7 cents? 8 cents? 9 cents? 10 cents?

25. If one pear cost 2 cents, what will $1\frac{1}{2}$ pears cost?

Ans. 3 cents; for if one pear cost 2 cents, $\frac{1}{2}$ pear will cost 1 cent, and 2 and 1 are 3.

26. What will $2\frac{1}{2}$ pears cost? $3\frac{1}{2}$? $4\frac{1}{2}$? $5\frac{1}{2}$? $6\frac{1}{2}$? $7\frac{1}{2}$? $8\frac{1}{2}$? $9\frac{1}{2}$? $10\frac{1}{2}$? $11\frac{1}{2}$? $12\frac{1}{2}$?

REM.—If one pear cost 3 cents, $\frac{1}{3}$ pear will cost 1 cent, and $\frac{2}{3}$ pear will cost twice as much as $\frac{1}{3}$, that is, 2 cents, and $\frac{4}{3} = 1\frac{1}{3}$ pears will cost 4 cents.

27. What part of a pear will 1 cent buy?   *Ans.* $\frac{1}{3}$.
28. What part of a pear will 2 cents buy?   *Ans.* $\frac{2}{3}$.
29. How many pears will 4 cents buy?

Ans. $\frac{4}{3} = 1\frac{1}{3}$ pears.

— ◆ —

LESSON XXIV.

At 3 cents apiece,

1. What cost $1\frac{2}{3}$ pears $= \frac{5}{3}$?   *Ans.* 5 cents.
2. How many can you buy for 5 cents?   *Ans.* $1\frac{2}{3}$.
3. What cost $2\frac{1}{3}$ pears $= \frac{7}{3}$?   *Ans.* 7 cents.
4. How many can you buy for 7 cents?   *Ans.* $2\frac{1}{3}$.
5. What cost $2\frac{2}{3}$ pears $= \frac{8}{3}$?   *Ans.* 8 cents.
6. How many can you buy for 8 cents?   *Ans.* $2\frac{2}{3}$.
7. What cost $3\frac{1}{3}$ pears $= \frac{10}{3}$?   *Ans.* 10 cents.
8. How many can you buy for 10 cents?   *Ans.* $3\frac{1}{3}$.
9. What cost $3\frac{2}{3}$ pears $= \frac{11}{3}$?   *Ans.* 11 cents.
10. How many can you buy for 11 cents?   *Ans.* $3\frac{2}{3}$.
11. What cost $4\frac{1}{3}$ pears $= \frac{13}{3}$?   *Ans.* 13 cents.
12. How many can you buy for 13 cents?   *Ans.* $4\frac{1}{3}$.
13. What cost $4\frac{2}{3}$ pears $= \frac{14}{3}$?   *Ans.* 14 cents.
14. How many can you buy for 14 cents?   *Ans.* $4\frac{2}{3}$.

LESSON XXV.

If one lemon cost 4 cents,

1. What will $\frac{1}{4}$ of a lemon cost?   *Ans.* 1 cent.
2. What part can you buy for 1 cent?   *Ans.* $\frac{1}{4}$.
3. What will $\frac{2}{4}$ of a lemon cost?   *Ans.* 2 cents.
4. What part can you buy for 2 cents?   *Ans.* $\frac{2}{4}$.
5. What will $\frac{3}{4}$ of a lemon cost?   *Ans.* 3 cents.
·6. What part can you buy for 3 cents?   *Ans.* $\frac{3}{4}$.
7. What will $\frac{5}{4} = 1\frac{1}{4}$ lemons cost?   *Ans.* 5 cents.
8. How many can you buy for 5 cents? *Ans.* $\frac{5}{4} = 1\frac{1}{4}$.
9. What will $\frac{6}{4} = 1\frac{2}{4}$ lemons cost?   *Ans.* 6 cents.
10. How many can you buy for 6 cents? *Ans.* $\frac{6}{4} = 1\frac{2}{4}$.
11. What will $\frac{7}{4} = 1\frac{3}{4}$ lemons cost?   *Ans.* 7 cents.
12. How many can you buy for 7 cents? *Ans.* $\frac{7}{4} = 1\frac{3}{4}$.
13. What will $\frac{8}{4} = 2$ lemons cost?   *Ans.* 8 cents.
14. How many can you buy for 8 cents?   *Ans.* $\frac{8}{4} = 2$.
15. What will $\frac{9}{4} = 2\frac{1}{4}$ lemons cost?   *Ans.* 9 cents.
16. How many can you buy for 9 cents?   *Ans.* $\frac{9}{4} = 2\frac{1}{4}$.
17. What will $\frac{10}{4} = 2\frac{2}{4}$ lemons cost?   *Ans.* 10 cents.
18. How many can you buy for 10 cents? *Ans.* $\frac{10}{4} = 2\frac{2}{4}$.
19. What will $\frac{11}{4} = 2\frac{3}{4}$ lemons cost?   *Ans.* 11 cents.
20. How many can you buy for 11 cents? *Ans.* $\frac{11}{4} = 2\frac{3}{4}$.
21. What will $\frac{12}{4} = 3$ lemons cost?   *Ans.* 12 cents.
22. How many can you buy for 12 cents? *Ans.* $\frac{12}{4} = 3$.
23. What will $\frac{13}{4} = 3\frac{1}{4}$ lemons cost?   *Ans.* 13 cents.
24. How many can you buy for 13 cents?   *Ans.* $3\frac{1}{4}$.
25. What will $\frac{14}{4} = 3\frac{2}{4}$ lemons cost?   *Ans.* 14 cents.
26. How many can you buy for 14 cents?   *Ans.* $3\frac{2}{4}$.

REM.—As every fourth costs 1 cent, the cost will be as many cents as there are fourths; and conversely, for every cent one-fourth of a pear can be bought.

LESSON XXVI.

If an orange cost 5 cents,

1. What will $\frac{1}{5}$ of an orange cost?   *Ans.* 1 cent.
2. What part will 1 cent buy?   *Ans.* $\frac{1}{5}$.
3. What will $\frac{2}{5}$ of an orange cost?   *Ans.* 2 cents.
4. What part will 2 cents buy?   *Ans.* $\frac{2}{5}$.
5. What will $\frac{3}{5}$ of an orange cost?   *Ans.* 3 cents.
6. What part will 3 cents buy?   *Ans.* $\frac{3}{5}$.
7. What will $\frac{4}{5}$ of an orange cost?   *Ans.* 4 cents.
8. What part will 4 cents buy?   *Ans.* $\frac{4}{5}$.
9. What will $\frac{5}{5}$ = 1 orange cost?   *Ans.* 5 cents.
10. How many will 5 cents buy?   *Ans.* $\frac{5}{5}$ = 1.
11. What will $\frac{6}{5}$ = $1\frac{1}{5}$ oranges cost?   *Ans.* 6 cents.
12. How many will 6 cents buy?   *Ans.* $\frac{6}{5}$ = $1\frac{1}{5}$.
13. What will $\frac{7}{5}$ = $1\frac{2}{5}$ oranges cost?   *Ans.* 7 cents.
14. How many will 7 cents buy?   *Ans.* $\frac{7}{5}$ = $1\frac{2}{5}$.
15. What will $\frac{8}{5}$ = $1\frac{3}{5}$ oranges cost?   *Ans.* 8 cents.
16. How many will 8 cents buy?   *Ans.* $\frac{8}{5}$ = $1\frac{3}{5}$.
17. What will $\frac{9}{5}$ = $1\frac{4}{5}$ oranges cost?   *Ans.* 9 cents.
18. How many will 9 cents buy?   *Ans.* $\frac{9}{5}$ = $1\frac{4}{5}$.
19. What will $\frac{10}{5}$ = 2 oranges cost?   *Ans.* 10 cents.
20. How many will 10 cents buy?   *Ans.* $\frac{10}{5}$ = 2.

Rᴇᴍ.—The results of all similar fractions will be similar.

———◆◆◆———

LESSON XXVII.

1. How many are 2 times $1\frac{1}{2}$? $2\frac{1}{2}$? $3\frac{1}{2}$? $4\frac{1}{2}$? $5\frac{1}{2}$? $6\frac{1}{2}$? $7\frac{1}{2}$? $8\frac{1}{2}$? $9\frac{1}{2}$? $10\frac{1}{2}$? $11\frac{1}{2}$? $12\frac{1}{2}$?

2. How many are 3 times $1\frac{1}{3}$? $1\frac{2}{3}$? $2\frac{1}{3}$? $2\frac{2}{3}$? $3\frac{1}{3}$? $3\frac{2}{3}$? $4\frac{1}{3}$? $4\frac{2}{3}$? $5\frac{1}{3}$? $5\frac{2}{3}$? $6\frac{1}{3}$? $6\frac{2}{3}$?

3. How many are 4 times $1\frac{1}{4}$? $1\frac{2}{4}$? $1\frac{3}{4}$? $2\frac{1}{4}$? $2\frac{2}{4}$? $2\frac{3}{4}$? $3\frac{1}{4}$? $3\frac{2}{4}$? $3\frac{3}{4}$? $4\frac{1}{4}$? $4\frac{2}{4}$? $4\frac{3}{4}$?

4. How many are 5 times $1\frac{1}{5}$? $1\frac{2}{5}$? $1\frac{3}{5}$? $1\frac{4}{5}$? $2\frac{1}{5}$? $2\frac{2}{5}$? $2\frac{3}{5}$? $2\frac{4}{5}$? $3\frac{1}{5}$? $3\frac{2}{5}$? $3\frac{3}{5}$? $3\frac{4}{5}$?

LESSON XXVIII.

1. How many times 2 are 3?
   Ans. Once two and one-half of two $= 1\frac{1}{2}$.
2. How many times 3 are 7? *Ans.* Twice 3 and $\frac{1}{3}$ of 3.
3. How many times 2 are 5? 7? 9? 11? 13? 15? 17? 19? 21? 23? 25?
4. How many times 3 are 4? 5? 7? 8? 10? 11? 13? 14? 16? 17? 19? 20?
5. How many times 4 are 5? 6? 7? 9? 10? 11? 13? 14? 15? 17? 18? 19?
6. How many times 5 are 6? 7? 8? 9? 11? 12? 13? 14? 16? 17? 18? 19?

LESSON XXIX.

1. What part of 2 is 1?
2. What part of 3 is 1? 2?
3. What part of 4 is 1? 2? 3?
4. What part of 5 is 1? 2? 3? 4?
5. What part of 6 is 1? 2? 3? 4? 5?
6. What part of 7 is 1? 2? 3? 4? 5? 6?
7. What part of 8 is 1? 2? 3? 4? 5? 6? 7?
8. What part of 9 is 1? 2? 3? 4? 5? 6? 7? 8?
9. What part of 10 is 1? 2? 3? 4? 5? 6? 7? 8? 9?

PRACTICAL EXAMPLES.

1. If a barrel of flour cost 8 dollars, what will $\frac{1}{8}$ of a barrel cost? $\frac{2}{8}$? $\frac{3}{8}$? $\frac{4}{8}$? $\frac{5}{8}$? $\frac{6}{8}$? $\frac{7}{8}$? $\frac{10}{8}$? $\frac{11}{8}$? $\frac{12}{8}$? $\frac{13}{8}$? $\frac{21}{8}$? In $\frac{24}{8}$ how many barrels? In $\frac{34}{8}$ how many barrels? In $\frac{31}{8}$ how many barrels? In $\frac{42}{8}$ how many barrels? What will $\frac{47}{8}$ cost?

2. If a barrel of flour cost 8 dollars, how many barrels can be bought for $13? $15? $17? $19? $21? $23? $24? $25? .$26? $31? $33? $34? $35? $37? $39? $40? $43? $45? $47? $49? $51? $53? $57? $61? $63? $65? $67? $69? $72?

3. If a ton of coal cost 6 dollars, what will $\frac{1}{6}$ of a ton cost? $\frac{2}{6}$? $\frac{3}{6}$? $\frac{4}{6}$? $\frac{5}{6}$? $\frac{7}{6}$? $\frac{8}{6}$? $\frac{12}{6}$? $\frac{21}{6}$? $\frac{30}{6}$? $\frac{32}{6}$? $\frac{40}{6}$? In $\frac{11}{6}$ how many tons? in $\frac{17}{6}$? $\frac{19}{6}$? $\frac{21}{6}$? $\frac{30}{6}$? $\frac{36}{6}$? $\frac{40}{6}$? $\frac{45}{6}$? $\frac{50}{6}$?

4. If a ton cost 6 dollars, how many tons can be bought for $10? for $12? $13? $15? $17? $19? $21? $27? $31? $35? $38? $40? $41? $45? $48?

REM.—The two questions, "What part of?" and "How many times?" are of the same import.

5. If a cord of wood cost 5 dollars, how many cords can be bought for $10? $11? $12? $13? $15? $20? 25? $26? $30? $31?

6. How many halves make a whole one? How many thirds? How many fourths? How many fifths? How many sixths? How many sevenths? How many tenths? How many fifteenths? How many fiftieths? How many hundredths? How many nine hundred and ninety-ninths? How many thousandths?

Ans.—Just as many as the number of parts into which the thing or number has been divided.

L E S S O N X X X.

1. 25 are how many times 4 ? 5 ? 6 ? 7 ? 8 ? 9 ?
10 ? 11 ? 12 ?

2. 30 are how many times 4 ? 5 ? 6 ? 7 ? 8 ? 9 ?
10 ? 11 ? 12 ?

3. 35 are how many times 4 ? 5 ? 6 ? 7 ? 8 ? 9 ?
10 ? 11 ? 12 ?

4. 36 are how many times 4 ? 5 ? 6 ? 7 ? 8 ? 9 ?
10 ? 11 ? 12 ?

5. 39 are how many times 4 ? 5 ? 6 ? 7 ? 8 ? 9 ?
10 ? 11 ? 12 ?

6. 40 are how many times 4 ? 5 ? 6 ? 7 ? 8 ? 9 ?
10 ? 11 ? 12 ?

7. 42 are how many times 4 ? 5 ? 6 ? 7 ? 8 ? 9 ?
10 ? 11 ? 12 ?

8. 44 are how many times 4 ? 5 ? 6 ? 7 ? 8 ? 9 ?
10 ? 11 ? 12 ?

9. 45 are how many times 4 ? 5 ? 6 ? 7 ? 8 ? 9 ?
10 ? 11 ? 12 ?

10. 46 are how many times 4 ? 5 ? 6 ? 7 ? 8 ? 9 ?
10 ? 11 ? 12 ?

11. 48 are how many times 4 ? 5 ? 6 ? 7 ? 8 ? 9 ?
10 ? 11 ? 12 ?

12. 50 are how many times 4 ? 5 ? 6 ? 7 ? 8 ? 9 ?
10 ? 11 ? 12 ?

13. 52 are how many times 4 ? 5 ? 6 ? 7 ? 8 ? 9 ?
10 ? 11 ? 12 ?

14. 54 are how many times 4 ? 5 ? 6 ? 7 ? 8 ? 9 ?
10 ? 11 ? 12 ?

15. 56 are how many times 4 ? 5 ? 6 ? 7 ? 8 ? 9 ?
10 ? 11 ? 12 ?

LESSON XXXI.

1. How many is one-half of 4? 6? 10? 12? 13? 14? 15? 16? 17? 18? 19? 20? 21? 22?

2. How many is one-third of 6? 8? 9? 12? 15? 18? 19? 20? 21? 22? 23? 24? 25? 26?

3. How many is one-fourth of 16? $\frac{2}{4}$ of 16? $\frac{3}{4}$? $\frac{5}{4}$? $\frac{6}{4}$? $\frac{7}{4}$? $\frac{9}{4}$? $\frac{10}{4}$? $\frac{11}{4}$? $\frac{12}{4}$? $\frac{13}{4}$? $\frac{14}{4}$?

4. How many is one-fifth of 15? $\frac{2}{5}$? $\frac{3}{5}$? $\frac{4}{5}$? $\frac{6}{5}$? $\frac{7}{5}$? $\frac{8}{5}$? $\frac{9}{5}$? $\frac{11}{5}$? $\frac{12}{5}$?

5. How many is one-sixth of 18? $\frac{2}{6}$? $\frac{3}{6}$? $\frac{4}{6}$? $\frac{5}{6}$? $\frac{7}{6}$? $\frac{8}{6}$? $\frac{9}{6}$? $\frac{10}{6}$? $\frac{11}{6}$?

6. How many is one-seventh of 21? $\frac{2}{7}$? $\frac{3}{7}$? $\frac{4}{7}$? $\frac{6}{7}$? $\frac{8}{7}$? $\frac{9}{7}$? $\frac{8}{7}$? $\frac{10}{7}$? $\frac{11}{7}$? $\frac{12}{7}$?

7. How many is one-eighth of 16? $\frac{2}{8}$? $\frac{3}{8}$? $\frac{4}{8}$? $\frac{5}{8}$? $\frac{6}{8}$? $\frac{7}{8}$? $\frac{9}{8}$? $\frac{10}{8}$? $\frac{11}{8}$? $\frac{12}{8}$?

8. How many is one-ninth of 18? $\frac{2}{9}$? $\frac{3}{9}$? $\frac{4}{9}$? $\frac{5}{9}$? $\frac{6}{9}$? $\frac{7}{9}$? $\frac{8}{9}$?

9. How many is one-tenth of 20? $\frac{2}{10}$? $\frac{3}{10}$? $\frac{4}{10}$? $\frac{5}{10}$? $\frac{6}{10}$? $\frac{7}{10}$? $\frac{8}{10}$? $\frac{9}{10}$?

PRACTICAL EXAMPLES.

1. If 5 apples cost 20 cents, what will 8 apples cost?

ANALYSIS.—One apple will cost $\frac{1}{5}$ as much as 5 apples, that is, $\frac{1}{5}$ of 20 cents, which is 4 cents; and 8 apples will cost eight times as much as 1 apple, that is, 8 times 4 cents, which is 32 cents.

2. If 3 pears cost 9 cents, what will 12 pears cost?

3. If 5 lemons cost 25 cents, what will 9 lemons cost?

4. If 5 oranges cost 30 cents, what will 10 oranges cost?

5. If 5 pounds of sugar cost 50 cents, what will 12 pounds cost?

6. If 5 yards of cloth cost 15 dollars, what will 20 yards cost ?

7. If 5 barrels of flour cost 30 dollars, what will 10 barrels cost ?

8. If 9 barrels of apples cost 27 dollars, what will 4 barrels cost ?

9. If 12 barrels of cider cost 36 dollars, what will 5 barrels cost ?

10. If 8 tons of coal cost 48 dollars, what will 5 tons cost ?

LESSON XXXII.

1. What is one-fourth of 4 ? 8 ? 12 ? 16 ? 20 ? 24 ? 28 ? 32 ? 36 ? 40 ? 44 ? 48 ?

2. What is $\frac{1}{5}$ of 5 ? 10 ? 15 ? 20 ? 25 ? 30 ? 35 ? 40 ? 45 ? 50 ? 55 ? 60 ?

3. What is $\frac{1}{6}$ of 6 ? 12 ? 18 ? 24 ? 30 ? 36 ? 42 ? 48 ? 54 ? 60 ? 66 ? 72 ?

4. What is $\frac{1}{7}$ of 7 ? 14 ? 21 ? 28 ? 35 ? 42 ? 49 ? 56 ? 63 ? 70 ? 77 ? 84 ?

5.. What is $\frac{1}{8}$ of 8 ? 16 ? 24 ? 32 ? 40 ? 48 ? 56 ? 64 ? 72 ? 80 ? 88 ? 96 ?

6. What is $\frac{1}{9}$ of 9 ? 18 ? 27 ? 36 ? 45 ? 54 ? 63 ? 72 ? 81 ? 90 ? 99 ? 108 ?

7. What is $\frac{1}{10}$ of 10 ? 20 ? 30 ? 40 ? 50 ? 60 ? 70 ? 80 ? 90 ? 100 ? 110 ? 120 ?

8. What is $\frac{1}{11}$ of 11 ? 22 ? 33 ? 44 ? 55 ? 66 ? 77 ? 88 ? 99 ? 110 ? 121 ? 132 ?

9. What is $\frac{1}{12}$ of 12 ? 24 ? 36 ? 48 ? 60 ? 72 ? 84 ? 96 ? 108 ? 120 ? 132 ? 144 ?

LESSON XXXIII.

1. One-half of 2 is $\frac{1}{3}$ of what number? $\frac{1}{4}$? $\frac{1}{5}$? $\frac{1}{6}$?
$\frac{1}{7}$? $\frac{1}{8}$? $\frac{1}{9}$? $\frac{1}{10}$? $\frac{1}{11}$? $\frac{1}{12}$?

2. One-half of 4 is $\frac{1}{3}$ of what number? $\frac{1}{4}$? $\frac{1}{5}$? $\frac{1}{6}$?
$\frac{1}{7}$? $\frac{1}{8}$? $\frac{1}{9}$? $\frac{1}{10}$? $\frac{1}{11}$? $\frac{1}{12}$?

3. One-half of 6 is $\frac{1}{3}$ of what number? $\frac{1}{4}$? $\frac{1}{5}$? $\frac{1}{6}$?
$\frac{1}{7}$? $\frac{1}{8}$? $\frac{1}{9}$? $\frac{1}{10}$? $\frac{1}{11}$? $\frac{1}{12}$?

4. One-half of 8 is $\frac{1}{3}$ of **what number?** $\frac{1}{4}$? $\frac{1}{5}$? $\frac{1}{6}$?
$\frac{1}{7}$? $\frac{1}{8}$? $\frac{1}{9}$? $\frac{1}{10}$? $\frac{1}{11}$? $\frac{1}{12}$?

5. One-half of 10 is $\frac{1}{3}$ of what number? $\frac{1}{4}$? $\frac{1}{5}$? $\frac{1}{6}$?
$\frac{1}{7}$? $\frac{1}{8}$? $\frac{1}{9}$? $\frac{1}{10}$? $\frac{1}{11}$? $\frac{1}{12}$?

6. One-half of 12 is $\frac{1}{3}$ of what number? $\frac{1}{4}$? $\frac{1}{5}$? $\frac{1}{6}$?
$\frac{1}{7}$? $\frac{1}{8}$? $\frac{1}{9}$? $\frac{1}{10}$? $\frac{1}{11}$? $\frac{1}{12}$?

7. One-third of 6 is $\frac{1}{2}$ of what number? $\frac{1}{4}$? $\frac{1}{5}$? $\frac{1}{6}$?
$\frac{1}{7}$? $\frac{1}{8}$? $\frac{1}{9}$? $\frac{1}{10}$? $\frac{1}{11}$? $\frac{1}{12}$?

8. One-third of 9 is $\frac{1}{2}$ of what number? $\frac{1}{4}$? $\frac{1}{5}$? $\frac{1}{6}$?
$\frac{1}{7}$? $\frac{1}{8}$? $\frac{1}{9}$? $\frac{1}{10}$? $\frac{1}{11}$? $\frac{1}{12}$?

9. One-third of 12 is $\frac{1}{2}$ of what number? $\frac{1}{4}$? $\frac{1}{5}$? $\frac{1}{6}$?
$\frac{1}{7}$? $\frac{1}{8}$? $\frac{1}{9}$? $\frac{1}{10}$? $\frac{1}{11}$? $\frac{1}{12}$?

10. One-third of 15 is $\frac{1}{2}$ of what number? $\frac{1}{4}$? $\frac{1}{5}$? $\frac{1}{6}$?
$\frac{1}{7}$? $\frac{1}{8}$? $\frac{1}{9}$? $\frac{1}{10}$? $\frac{1}{11}$? $\frac{1}{12}$?

11. One-third of 18 is $\frac{1}{2}$ of what number? $\frac{1}{4}$? $\frac{1}{5}$? $\frac{1}{6}$?
$\frac{1}{7}$? $\frac{1}{8}$? $\frac{1}{9}$? $\frac{1}{10}$? $\frac{1}{11}$? $\frac{1}{12}$?

12. One-fourth of 8 is $\frac{1}{2}$ of what number? $\frac{1}{4}$? $\frac{1}{5}$? $\frac{1}{6}$?
$\frac{1}{7}$? $\frac{1}{8}$? $\frac{1}{9}$? $\frac{1}{10}$? $\frac{1}{11}$? $\frac{1}{12}$?

13. One-fourth of 12 is $\frac{1}{2}$ of what number? $\frac{1}{3}$? $\frac{1}{5}$? $\frac{1}{6}$?
$\frac{1}{7}$? $\frac{1}{8}$? $\frac{1}{9}$? $\frac{1}{10}$? $\frac{1}{11}$? $\frac{1}{12}$?

14. One-fourth of 16 is $\frac{1}{2}$ of what number? $\frac{1}{3}$? $\frac{1}{5}$? $\frac{1}{6}$?
$\frac{1}{7}$? $\frac{1}{8}$? $\frac{1}{9}$? $\frac{1}{10}$? $\frac{1}{11}$? $\frac{1}{12}$?

15. One-fourth of 20 is $\frac{1}{2}$ of what number? $\frac{1}{3}$? $\frac{1}{5}$? $\frac{1}{6}$?
$\frac{1}{7}$? $\frac{1}{8}$? $\frac{1}{9}$? $\frac{1}{10}$? $\frac{1}{11}$? $\frac{1}{12}$?

16. One-fourth of 24 is $\frac{1}{2}$ of what number? $\frac{1}{3}$? $\frac{1}{4}$? $\frac{1}{6}$? $\frac{1}{7}$? $\frac{1}{8}$? $\frac{1}{9}$? $\frac{1}{10}$? $\frac{1}{11}$? $\frac{1}{12}$?

17. One-fifth of 10 is $\frac{1}{2}$ of what number? $\frac{1}{3}$? $\frac{1}{4}$? $\frac{1}{6}$? $\frac{1}{7}$? $\frac{1}{8}$? $\frac{1}{9}$? $\frac{1}{10}$? $\frac{1}{11}$? $\frac{1}{12}$?

18. One-fifth of 15 is $\frac{1}{2}$ of what number? $\frac{1}{3}$? $\frac{1}{4}$? $\frac{1}{6}$? $\frac{1}{7}$? $\frac{1}{8}$? $\frac{1}{9}$? $\frac{1}{10}$? $\frac{1}{11}$? $\frac{1}{12}$?

19. One-fifth of 20 is $\frac{1}{2}$ of what number? $\frac{1}{3}$? $\frac{1}{4}$? $\frac{1}{6}$? $\frac{1}{7}$? $\frac{1}{8}$? $\frac{1}{9}$? $\frac{1}{10}$? $\frac{1}{11}$? $\frac{1}{12}$?

20. One-fifth of 25 is $\frac{1}{2}$ of what number? $\frac{1}{3}$? $\frac{1}{4}$? $\frac{1}{6}$? $\frac{1}{7}$? $\frac{1}{8}$? $\frac{1}{9}$? $\frac{1}{10}$? $\frac{1}{11}$? $\frac{1}{12}$?

21. One-fifth of 30 is $\frac{1}{2}$ of what number? $\frac{1}{3}$? $\frac{1}{4}$? $\frac{1}{6}$? $\frac{1}{7}$? $\frac{1}{8}$? $\frac{1}{9}$? $\frac{1}{10}$? $\frac{1}{11}$? $\frac{1}{12}$?

22. One-sixth of 12 is $\frac{1}{2}$ of what number? $\frac{1}{3}$? $\frac{1}{4}$? $\frac{1}{5}$? $\frac{1}{7}$? $\frac{1}{8}$? $\frac{1}{9}$? $\frac{1}{10}$? $\frac{1}{11}$? $\frac{1}{12}$?

23. One-sixth of 18 is $\frac{1}{2}$ of what number? $\frac{1}{3}$? $\frac{1}{4}$? $\frac{1}{5}$? $\frac{1}{7}$? $\frac{1}{8}$? $\frac{1}{9}$? $\frac{1}{10}$? $\frac{1}{11}$? $\frac{1}{12}$?

24. One-sixth of 24 is $\frac{1}{2}$ of what number? $\frac{1}{3}$? $\frac{1}{4}$? $\frac{1}{5}$? $\frac{1}{7}$? $\frac{1}{8}$? $\frac{1}{9}$? $\frac{1}{10}$? $\frac{1}{11}$? $\frac{1}{12}$?

25. One-sixth of 30 is $\frac{1}{2}$ of what number? $\frac{1}{3}$? $\frac{1}{4}$? $\frac{1}{5}$? $\frac{1}{7}$? $\frac{1}{8}$? $\frac{1}{9}$? $\frac{1}{10}$? $\frac{1}{11}$? $\frac{1}{12}$?

26. One-sixth of 36 is $\frac{1}{2}$ of what number? $\frac{1}{3}$? $\frac{1}{4}$? $\frac{1}{5}$? $\frac{1}{7}$? $\frac{1}{8}$? $\frac{1}{9}$? $\frac{1}{10}$? $\frac{1}{11}$? $\frac{1}{12}$?

---◆•◆---

LESSON XXXIV.

1. In 1 peck there are 8 quarts; how many quarts in 2 pecks? 3? 4? 5? 6? 7? 8? 9? 10? 11? 12?

2. How many pecks in 8 quarts? in 16 quarts? 24? 32? 40? 48? 56? 64? 72? 80? 88? 96?

3. In 1 bushel there are 4 pecks; how many pecks in 2 bushels? in 3 bushels? 4? 5? 6? 7? 8? 9? 10? 11? 12?

4. How many bushels in 4 pecks? in 8 pecks? 12? 16? 20? 24? 28? 32? 36? 40? 44? 48?

5. At 2 cents a pound, what will $3\frac{1}{2}$ pounds of flour cost? 3 times 2 are 6, and one-half of 2 is 1, and 6 and 1 are 7; therefore, 3 times 2 and one-half of 2 are 7.

6. What will $4\frac{1}{2}$ pounds cost? $5\frac{1}{2}$? $6\frac{1}{2}$? $7\frac{1}{2}$? $8\frac{1}{2}$? $9\frac{1}{2}$? $10\frac{1}{2}$? $11\frac{1}{2}$? $12\frac{1}{2}$?

7. At 3 cents a pound, what cost $1\frac{1}{3}$ pounds? Once 3 is 3 and $\frac{1}{3}$ of 3 is 1; 3 and 1 are 4. Once three and $\frac{1}{3}$ of 3 are 4.

8. What cost $1\frac{2}{3}$ pounds? $2\frac{1}{3}$? $2\frac{2}{3}$? $3\frac{1}{3}$? $3\frac{2}{3}$? $4\frac{1}{3}$? $4\frac{2}{3}$? $5\frac{1}{3}$? $5\frac{2}{3}$? $6\frac{1}{3}$? $6\frac{2}{3}$? $7\frac{1}{3}$? $7\frac{2}{3}$? $8\frac{1}{3}$? $8\frac{2}{3}$? $9\frac{1}{3}$? $9\frac{2}{3}$? $10\frac{1}{3}$? $10\frac{2}{3}$? $11\frac{1}{3}$? $11\frac{2}{3}$?

9. How many are 3 times 4 and one-fourth of 4? 3 times 4 and $\frac{2}{4}$ of 4? 4 times 4 and $\frac{3}{4}$ of 4? 5 times 4 and $\frac{1}{4}$ of 4? 6 times 4 and $\frac{1}{4}$ of 4? 7 times 4 and $\frac{2}{4}$ of 4? 8 times 4 and $\frac{3}{4}$ of 4? 9 times 4 and $\frac{1}{4}$ of 4? 10 times 4 and $\frac{2}{4}$ of 4? 11 times 4 and $\frac{1}{4}$ of 4? 12 times 4 and $\frac{3}{4}$ of 4?

10. How many are 3 times 5 and $\frac{1}{5}$ of 5? 3 times 5 and $\frac{2}{5}$ of 5? 4 times 5 and $\frac{1}{5}$ of 5? 4 times 5 and $\frac{4}{5}$ of 5? 5 times 5 and $\frac{3}{5}$ of 5? 5 times 5 and $\frac{1}{5}$ of 5? 6 times 5 and $\frac{2}{5}$ of 5? 6 times 5 and $\frac{4}{5}$ of 5? 7 times 5 and $\frac{2}{5}$ of 5? 7 times 5 and $\frac{3}{5}$ of 5? 8 times 5 and $\frac{2}{5}$ of 5? 8 times 5 and $\frac{4}{5}$ of 5? 9 times 5 and $\frac{1}{5}$ of 5? 9 times 5 and $\frac{3}{5}$ of 5? 10 times 5 and $\frac{2}{5}$ of 5? 11 times 5 and $\frac{1}{5}$ of 5? 12 times 5 and $\frac{4}{5}$ of 5?

LESSON XXXV.

1. 64 are how many times 6 ? 7 ? 8 ? 9 ? 10 ? 11 ? 12 ?

2. 66 are how many times 6 ? 7 ? 8 ? 9 ? 10 ? 11 ? 12 ?

3. 70 are how many times 6 ? 7 ? 8 ? 9 ? 10 ? 11 ? 12 ?

4. 72 are how many times 6 ? 7 ? 8 ? 9 ? 10 ? 11 ? 12 ?

5. 75 are how many times 6 ? 7 ? 8 ? 9 ? 10 ? 11 ? 12 ?

6. 76 are how many times 6 ? 7 ? 8 ? 9 ? 10 ? 11 ? 12 ?

7. 80 are how many times 6 ? 7 ? 8 ? 9 ? 10 ? 11 ? 12 ?

8. 81 are how many times 6 ? 7 ? 8 ? 9 ? 10 ? 11 ? 12 ?

9. 82 are how many times 6 ? 7 ? 8 ? 9 ? 10 ? 11 ? 12 ?

10. 85 are how many times 6 ? 7 ? 8 ? 9 ? 10 ? 11 ? 12 ?

11. 90 are how many times 6 ? 7 ? 8 ? 9 ? 10 ? 11 ? 12 ?

12. 95 are how many times 6 ? 7 ? 8 ? 9 ? 10 ? 11 ? 12 ?

13. 100 are how many times 6 ? 7 ? 8 ? 9 ? 10 ? 11 ? 12 ?

14. 120 are how many times 6 ? 7 ? 8 ? 9 ? 10 ? 11 ? 12 ?

15. 144 are how many times 6 ? 7 ? 8 ? 9 ? 10 ? 11 ? 12 ?

3

LESSON XXXVI.

1. 2 is one-half of what number ? $\frac{1}{3}$ of what ? $\frac{1}{4}$? $\frac{1}{5}$? $\frac{1}{6}$? $\frac{1}{7}$? $\frac{1}{8}$? $\frac{1}{9}$? $\frac{1}{10}$? $\frac{1}{11}$? $\frac{1}{12}$?

2. 3 is $\frac{1}{2}$ of what number ? $\frac{1}{3}$ of what? $\frac{1}{4}$? $\frac{1}{5}$? $\frac{1}{6}$? $\frac{1}{7}$? $\frac{1}{8}$? $\frac{1}{9}$? $\frac{1}{10}$? $\frac{1}{11}$? $\frac{1}{12}$?

3. 4 is $\frac{1}{2}$ of what number ? $\frac{1}{3}$ of what? $\frac{1}{4}$? $\frac{1}{5}$? $\frac{1}{6}$? $\frac{1}{7}$? $\frac{1}{8}$? $\frac{1}{9}$? $\frac{1}{10}$? $\frac{1}{11}$? $\frac{1}{12}$?

4. 5 is $\frac{1}{2}$ of what number ? $\frac{1}{3}$ of what? $\frac{1}{4}$? $\frac{1}{5}$? $\frac{1}{6}$? $\frac{1}{7}$? $\frac{1}{8}$? $\frac{1}{9}$? $\frac{1}{10}$? $\frac{1}{11}$? $\frac{1}{12}$?

5. 6 is $\frac{1}{2}$ of what number ? $\frac{1}{3}$ of what? $\frac{1}{4}$? $\frac{1}{5}$? $\frac{1}{6}$? $\frac{1}{7}$? $\frac{1}{8}$? $\frac{1}{9}$? $\frac{1}{10}$? $\frac{1}{11}$? $\frac{1}{12}$?

6. 7 is $\frac{1}{2}$ of what number ? $\frac{1}{3}$ of what? $\frac{1}{4}$? $\frac{1}{5}$? $\frac{1}{6}$? $\frac{1}{7}$? $\frac{1}{8}$? $\frac{1}{9}$? $\frac{1}{10}$? $\frac{1}{11}$? $\frac{1}{12}$?

7. 8 is $\frac{1}{2}$ of what number ? $\frac{1}{3}$ of what? $\frac{1}{4}$? $\frac{1}{5}$? $\frac{1}{6}$? $\frac{1}{7}$? $\frac{1}{8}$? $\frac{1}{9}$? $\frac{1}{10}$? $\frac{1}{11}$? $\frac{1}{12}$?

8. 9 is $\frac{1}{2}$ of what number ? $\frac{1}{3}$ of what? $\frac{1}{4}$? $\frac{1}{5}$? $\frac{1}{6}$? $\frac{1}{7}$? $\frac{1}{8}$? $\frac{1}{9}$? $\frac{1}{10}$? $\frac{1}{11}$? $\frac{1}{12}$?

9. 10 is $\frac{1}{2}$ of what number ? $\frac{1}{3}$ of what? $\frac{1}{4}$? $\frac{1}{5}$? $\frac{1}{6}$? $\frac{1}{7}$? $\frac{1}{8}$? $\frac{1}{9}$? $\frac{1}{10}$? $\frac{1}{11}$? $\frac{1}{12}$?

10. 11 is $\frac{1}{2}$ of what number ? $\frac{1}{3}$ of what? $\frac{1}{4}$? $\frac{1}{5}$? $\frac{1}{6}$? $\frac{1}{7}$? $\frac{1}{8}$? $\frac{1}{9}$? $\frac{1}{10}$? $\frac{1}{11}$? $\frac{1}{12}$?

11. 12 is $\frac{1}{2}$ of what number ? $\frac{1}{3}$ of what? $\frac{1}{4}$? $\frac{1}{5}$? $\frac{1}{6}$? $\frac{1}{7}$? $\frac{1}{8}$? $\frac{1}{9}$? $\frac{1}{10}$? $\frac{1}{11}$? $\frac{1}{12}$?

LESSON XXXVII.

1. 4 is $\frac{2}{3}$ of what number ?

Ans., If 4 is $\frac{2}{3}$ of some number, $\frac{1}{2}$ of 4 = 2 is $\frac{1}{3}$ of the same number, 2 is $\frac{1}{3}$ of 6.

2. 6 is $\frac{3}{4}$ of what number?

Ans., If 6 is $\frac{3}{4}$ of some number, $\frac{1}{3}$ of 6 = 2 is $\frac{1}{4}$ of the same number, and 2 is $\frac{1}{4}$ of 8.

3. 8 is $\frac{4}{5}$ of what number?

4. 9 is $\frac{3}{7}$ of what number?

5. 10 is $\frac{5}{8}$ of what number?

6. 12 is $\frac{6}{9}$ of what number?

7. 14 is $\frac{7}{10}$ of what number?

8. 16 is $\frac{8}{11}$ of what number?

9. 18 is $\frac{9}{12}$ of what number?

10. 20 is $\frac{10}{13}$ of what number?

11. What is $\frac{1}{4}$ of 4? $\frac{2}{4}$ of 4? $\frac{3}{4}$ of 4?

12. What is $\frac{1}{2}$ of 6? $\frac{1}{3}$ of 6? $\frac{1}{6}$ of 6?

13. What is $\frac{1}{2}$ of 8? $\frac{1}{4}$ of 8? $\frac{1}{8}$ of 8?

14. What is $\frac{1}{3}$ of 9? $\frac{1}{9}$ of 9? $\frac{2}{3}$ of 9? $\frac{1}{9}$ of 9?

15. What is $\frac{1}{2}$ of 10? $\frac{1}{5}$ of 10? $\frac{1}{4}$ of 10? $\frac{1}{8}$ of 10?

16. What is $\frac{1}{2}$ of 12? $\frac{1}{3}$ of 12? $\frac{1}{4}$ of 12? $\frac{1}{6}$ of 12?

17. What is $\frac{1}{2}$ of 14? $\frac{1}{3}$ of 14? $\frac{1}{4}$ of 14? $\frac{1}{7}$ of 14?

18. What is $\frac{1}{2}$ of 16? $\frac{1}{4}$ of 16? $\frac{1}{8}$ of 16? $\frac{1}{10}$ of 16?

19. What is $\frac{1}{2}$ of 18? $\frac{1}{3}$? $\frac{2}{3}$? $\frac{1}{6}$? $\frac{2}{6}$? $\frac{1}{9}$?

20. What is $\frac{1}{2}$ of 20? $\frac{1}{4}$? $\frac{3}{4}$? $\frac{1}{5}$? $\frac{3}{5}$? $\frac{1}{10}$? $\frac{4}{10}$?

PRACTICAL EXAMPLES.

1. If 2 yards of cloth cost 6 dollars, what will 9 yards cost?

2. If 3 yards of cloth cost 9 dollars, what will 5 yards cost?

3. If 4 yards of cloth cost 6 dollars, what will 12 yards cost?

4. If 5 yards of cloth cost 15 dollars, what will 8 yards cost?

5. If 6 firkins of butter cost 30 dollars, what will 11 firkins cost?

6. If 7 firkins of butter cost 42 dollars, what will 12 firkins cost?

7. If 8 firkins of butter cost 40 dollars, what will 3 firkins cost?

8. If 9 barrels of flour cost 54 dollars, what will 4 barrels cost?

9. If 10 barrels of flour cost 70 dollars, what will 2 barrels cost?

10. If 11 barrels of flour cost 66 dollars, what will 6 barrels cost?

11. If 12 barrels of flour cost 96 dollars, what will 1 barrel cost?

LESSON XXXVIII.

1. How many halves in $1\frac{1}{2}$? $2\frac{1}{2}$? $3\frac{1}{2}$? $4\frac{1}{2}$? $5\frac{1}{2}$? $6\frac{1}{2}$? $7\frac{1}{2}$? $8\frac{1}{2}$? $9\frac{1}{2}$? $10\frac{1}{2}$? $11\frac{1}{2}$? $12\frac{1}{2}$?

2. How many thirds in $1\frac{1}{3}$? $1\frac{2}{3}$? $2\frac{1}{3}$? $2\frac{2}{3}$? $3\frac{1}{3}$? $3\frac{2}{3}$? $4\frac{1}{3}$? $4\frac{2}{3}$? $5\frac{1}{3}$? $5\frac{2}{3}$?

3. How many fourths in $1\frac{1}{4}$? $1\frac{2}{4}$? $1\frac{3}{4}$? $2\frac{1}{4}$? $2\frac{2}{4}$? $2\frac{3}{4}$? $3\frac{1}{4}$? $3\frac{2}{4}$? $3\frac{3}{4}$? $4\frac{1}{4}$? $4\frac{2}{4}$? $4\frac{3}{4}$?

4. How many fifths in $1\frac{1}{5}$? $1\frac{2}{5}$? $1\frac{3}{5}$? $1\frac{4}{5}$? $2\frac{1}{5}$? $2\frac{2}{5}$? $2\frac{3}{5}$? $2\frac{4}{5}$? $3\frac{1}{5}$? $3\frac{2}{5}$? $3\frac{3}{5}$? $3\frac{4}{5}$?

5. How many sixths in $1\frac{1}{6}$? $1\frac{2}{6}$? $1\frac{3}{6}$? $1\frac{4}{6}$? $1\frac{5}{6}$? $2\frac{1}{6}$? $2\frac{3}{6}$? $3\frac{2}{6}$? $3\frac{5}{6}$? $4\frac{3}{6}$? $5\frac{1}{6}$?

REM.—In every unit there are 2 halves, 3 thirds, 4 fourths, 5 fifths, and 6 sixths.

COR.—When fractions have a common denominator, the quotient of their numerators is the quotient of the fractions.

The foregoing questions are the same as if the questions were one-half, one-third, etc., is contained in $1\frac{1}{2}$, $1\frac{1}{3}$, etc., how many times.

6. $\frac{1}{2}$ is contained in $1\frac{1}{2}$ how many times?

$$1\frac{1}{2} = \frac{3}{2}. \qquad \therefore \quad \frac{3}{2} \div \frac{1}{2} = 3.$$

Reduce both to a common denominator; the question then is, How often is 1 contained in 3? *Ans.* 3 times.

7. How often is $\frac{1}{3}$ contained in $1\frac{1}{3}$?

$$1\frac{1}{3} = \frac{4}{3}, \quad \text{and} \quad \frac{4}{3} \div \frac{1}{3} = 4, \text{ Ans.}$$

8. How often is $\frac{1}{4}$ contained in $1\frac{3}{4}$?

$$1\frac{3}{4} = \frac{7}{4}, \quad \text{and} \quad \frac{7}{4} \div \frac{1}{4} = 7, \text{ Ans.}$$

9. How often is $\frac{1}{7}$ contained in $\frac{2}{3}$?

Common denominator, 21.

$$\frac{2 \times 7}{3 \times 7} = \frac{14}{21}, \quad \text{and} \quad \frac{1 \times 3}{7 \times 3} = \frac{3}{21}.$$

$$14 \div 3 = 4\frac{2}{3}.$$

10. How often is $\frac{2}{3}$ contained in $3\frac{3}{4}$?

Common denominator, 12.

$$3\frac{3}{4} = \frac{15 \times 3}{4 \times 3} = \frac{45}{12}, \quad \text{and} \quad \frac{2 \times 4}{3 \times 4} = \frac{8}{12}.$$

$$45 \div 8 = 5\frac{5}{8}.$$

11. Divide $4\frac{6}{7}$ by $\frac{7}{8}$.

$$4\frac{6}{7} = \frac{34 \times 8}{7 \times 8} = \frac{272}{56}, \quad \text{and} \quad \frac{7 \times 7}{8 \times 7} = \frac{49}{56}.$$

$$272 \div 49 = 5\frac{27}{49}.$$

Reduce dividend and divisor to a common denominator, and use their numerators as integers.

$$4\frac{6}{7} = \frac{34}{7} \times \frac{8}{7} = \frac{272}{49} = 5\frac{27}{49}.$$

LESSON XXXIX.

1. 3 times $\frac{4}{5}$ are how many fifths? how many units?

2. 4 times $\frac{5}{6}$ are how many sixths? how many units?

3. 5 times $\frac{6}{7}$ are how many sevenths? how many units?

4. 6 times $\frac{7}{8}$ are how many eighths? how many units?

5. 7 times $\frac{8}{9}$ are how many ninths? how many units?

6. 8 times $\frac{9}{10}$ are how many tenths? how many units?

7. 9 times $\frac{10}{11}$ are how many elevenths? how many units?

8. 10 times $\frac{11}{12}$ are how many twelfths? how many units?

9. If a yard of cloth cost 3 dollars, what will $\frac{1}{4}$ yard cost? $\frac{2}{4}$? $\frac{3}{4}$? $\frac{4}{4}$? $\frac{7}{4}$? $\frac{8}{4}$? $\frac{10}{4}$?

10. If a yard of cloth cost 3 dollars, what will $\frac{1}{5}$ yard cost? $\frac{2}{5}$? $\frac{4}{5}$? $\frac{6}{5}$? $\frac{7}{5}$? $\frac{8}{5}$? $\frac{9}{5}$?

11. What is one-seventh of 1? of 2? 3? 4? 5? 6? 7? 8? 9? 10? 11? 12?

12. What is one-eighth of 1? of 2? 3? 4? 5? 6? 7? 8? 9? 10? 11? 12?

13. What is one-ninth of 1? of 2? 3? 4? 5? 6? 7? 8? 9? 10? 11? 12?

14. What is one-tenth of 1? of 2? 3? 4? 5? 6? 7? 8? 9? 10? 11? 12?

15. What is one-eleventh of 1? of 2? 3? 4? 5? 6? 7? 8? 9? 10? 11? 12?

16. What is one-twelfth of 1? of 2? 3? 4? 5? 6? 7? 8? 9? 10? 11? 12?

REM.—In all these examples, the $\frac{1}{2}$, $\frac{1}{3}$, $\frac{1}{4}$, etc., is contained in the dividend just as often as the numerator 1 of the divisor is contained in the numerator of the dividend. As the divisor and dividend have the same denominators, and are therefore like numbers, therefore, to divide a fraction by a fraction, reduce both to a common denominator, and divide the numerator of the dividend by the numerator of the divisor.

As they have a common denominator, their numerators are like integers.

17. Divide $\frac{5}{6}$ by $\frac{1}{2}$.

$$\tfrac{1}{2} = \tfrac{3}{6}. \qquad \therefore \quad 5 \div 3 = \tfrac{5}{3} = 1\tfrac{2}{3}.$$

18. Divide $4\tfrac{3}{5}$ by $\tfrac{2}{3}$.

Common denominator, 15.

$$4\tfrac{3}{5} = \frac{23 \times 3}{5 \times 3} = \frac{69}{15}, \quad \text{and} \quad \frac{2 \times 5}{3 \times 5} = \frac{10}{15}.$$

$$23 \div 10 = \tfrac{23}{10} = 2\tfrac{3}{10}.$$

19. Divide $\tfrac{2}{3}$ by $\tfrac{3}{7}$.

$$\frac{2 \times 7}{3 \times 7} = \frac{14}{21}, \quad \text{and} \quad \frac{3 \times 3}{7 \times 3} = \frac{9}{21}.$$

$$14 \div 9 = \tfrac{14}{9} = 1\tfrac{5}{9}.$$

The numerator of the quotient is the product of the present numerator and the denominator of the divisor, and the denominator of the quotient is the product of the numerator of the present divisor and the denominator of the dividend; thus,

$$\tfrac{2}{3} \times \tfrac{7}{3} = \tfrac{14}{9} = 1\tfrac{5}{9}.$$

20. Divide $25\tfrac{4}{9}$ by $\tfrac{3}{7}$.

$$25\tfrac{4}{9} = \frac{229 \times 7}{9 \times 7} = \frac{1603}{63}, \quad \text{and} \quad \frac{3 \times 9}{7 \times 9} = \frac{27}{63}.$$

If, therefore, the divisor be inverted and made a multiplier, the proper result will be obtained; for, in reducing two fractions to a common denominator, both terms of each fraction is multiplied by the denominator of the other fraction; and, as in division we want the divisor to be the denominator, it must therefore be inverted and then made a multiplier.

TABLES

ORDERS OF ABSTRACT NUMBERS.

10 units	= 1 ten.
10 tens	= 1 hundred.
10 hundreds	= 1 thousand.
10 thousands	= 1 ten-thousand.
10 ten-thousands	= 1 hundred-thousand.
10 hundred-thousands	= 1 million.
etc.	etc.

UNITED STATES MONEY.

10 mills (*m.*)	= 1 cent (*ct.*).
10 cents	= 1 dime.
10 dimes	= 1 dollar.
10 dollars	= 1 eagle.

Dimes and eagles are coins, but are not regarded in computation. 100 cents make 1 dollar.

ENGLISH MONEY.

4 farthings (*far.*)	= 1 penny (*d.*).
12 pence	= 1 shilling (*s.*).
20 shillings	= 1 pound (£).
21 shillings	= 1 guinea.

AVOIRDUPOIS WEIGHT.

Used in commercial transactions, when goods are bought or sold in quantity, and for all metals except gold and silver.

$$16 \text{ drams } (dr.) = 1 \text{ ounce } (oz.)$$
$$16 \text{ ounces} = 1 \text{ pound } (lb.).$$
$$25 \text{ pounds} = 1 \text{ quarter } (qr.).$$
$$4 \text{ quarters} = 1 \text{ hundredweight } (cwt.).$$
$$20 \text{ cwt.} = 1 \text{ ton } (T.).$$

TROY WEIGHT.

Used for gold, silver, and jewels; also in philosophical experiments.

$$24 \text{ grains } (gr.) = 1 \text{ pennyweight } (pwt.).$$
$$20 \text{ pennyweights} = 1 \text{ ounce.}$$
$$12 \text{ ounces} = 1 \text{ pound.}$$

DIAMOND WEIGHT.

Used for diamonds and other precious stones.

$$16 \text{ parts } = 1 \text{ grain } = .8 \text{ grain Troy.}$$
$$4 \text{ grains } = 1 \text{ carat } = 3.2 \text{ grains Troy.}$$

APOTHECARIES' WEIGHT.

Used by druggists in putting up prescriptions, the pound, ounce, and grain are the same as in Troy Weight.

$$20 \text{ grains } = 1 \text{ scruple } (\Im).$$
$$3 \text{ scruples } = 1 \text{ dram } (\mathfrak{Z}).$$
$$8 \text{ drams } = 1 \text{ ounce } (\mathfrak{Z}).$$
$$12 \text{ ounces } = 1 \text{ pound.}$$

Comparison of Weights.

1 pound Avoirdupois = 7000 grains Troy.
1 pound Troy   = 5760 grains Troy.

APOTHECARIES' FLUID WEIGHT.

Used for liquids in medical prescriptions.

60 minims (M) = 1 fluid dram (*f.* 3).
8 fluid drams   = 1 fluid ounce (*f.* $\mathfrak{Z}$).
16 fluid ounces = 1 pint (*O.*).
8 O.   = 1 gallon (*Cong.*)

For ordinary use, 1 teacup = 2 wine glasses =
8 tablespoons = 32 teaspoons = 4 *f.* $\mathfrak{Z}$.

LINEAR MEASURE.

Used for lengths and distances.

12 inches (*in.*)   = 1 foot (*ft.*).
3 feet   = 1 yard (*yd.*).
5½ yds. or 16½ ft. = 1 rod (*rd.*).
40 rods   = 1 furlong (*fur.*).
8 furlongs   = 1 mile (*mi.*)
3 miles   = 1 league (*lea.*).

CLOTH MEASURE.

2¼ inches   = 1 nail (*na.*).
4 na. or 9 in. = 1 quarter (*qr.*).
4 quarters   = 1 yard.
3 quarters   = 1 ell Flemish.
5 quarters   = 1 ell English.
6 quarters   = 1 ell French.

SURVEYORS' MEASURE OF LENGTH.

$7\frac{92}{100}$ in.	$= 1$ link ($l.$).
25 links	$= 1$ pole ($p.$).
100 l., 4 p., or 66 ft.	$= 1$ chain ($ch.$).
10 chains	$= 1$ furlong.
8 fur. $= 80$ ch.	$= 1$ mile.

MARINERS' MEASURE.

6 feet	$= 1$ fathom.
120 fathoms	$= 1$ cable length.
880 fathoms, or $7\frac{1}{3}$ cable lengths	$= 1$ mile.

REM. 1 nautical league equals 3 equatorial miles $= 3.45771$ statute miles.

60 equatorial miles equals 69.1542 statute miles $= 1$ equatorial degree.

$360°$ $=$ the circumference of a circle.

360 equatorial degrees $=$ the circumference of the earth.

LINEAR MEASURE.

12 inches	$= 1$ foot.
3 feet	$= 1$ yard.
$5\frac{1}{2}$ yards	$= 1$ rod, pole, or perch.

SQUARE MEASURE.

$12 \times 12 = 144$	sq. in.	$= 1$ sq. ft.
$3 \times 3 = 9$	sq. ft.	$= 1$ sq. yd.
$5\frac{1}{2} \times 5\frac{1}{2} = 30\frac{1}{4}$	sq. yd.	$= 1$ sq. rd. or perch.

CUBIC MEASURE.

$12 \times 12 \times 12 = 1728$ cu. in. $= 1$ cu. ft.
$3 \times 3 \times 3 = 27$ cu. ft. $= 1$ cu. yd.
16 cu. ft. $= 1$ cord foot.
8 cu. ft. $= 1$ cord wood, bark, etc.

REM. 40 cu. ft. round timber, or 50 ft. hewn timber $= 1$ ton.

A perch of stone is 16½ ft. long, 1½ ft. wide, and 1 ft. high $=$ 24½ cu. ft. This measure is caused by walls being generally 1½ ft. wide.

LAND MEASURE.

40 perches ($p.$) $= 1$ rood ($R.$).
4 roods $= 1$ acre ($A.$).
640 acres $= 1$ square mile, termed a section.

LIQUID MEASURE.

4 gills ($gi.$) $= 1$ pint ($pt.$).
2 pints $= 1$ quart ($qt.$).
4 quarts $= 1$ gallon ($gal.$).

REM.—In all liquids except ale, beer, and milk, the gallon is 231 cu. in. In ale, beer, and milk it is 282 cu. in. In the former, 31½ gallons is called a barrel; 63 gallons, a hogshead; 42 gallons, a tierce; 84 gallons, a puncheon; 126 gallons, a pipe; and 2 pipes, a tun. In the latter, 36 gallons $= 1$ barrel, and 54 gallons, a hogshead; these are only vessels, not measures.

DRY MEASURE.

Used for grain, fruit, vegetables, coal, salt, etc.

2 pints $= 1$ quart.
8 quarts $= 1$ peck ($pk.$).
4 pecks $= 1$ bushel ($bu.$).

The wine gallon of United States	=	231	cu. in.
The beer gallon of United States	=	282	cu. in.
The dry gallon of United States	=	268.8	cu. in.
Imperial gal. of Great Britain for dry and liquid measures	=	277.274	cu. in.
United States dry bushel	=	2150.42	cu. in.
Imperial bushel of Great Britain	=	2218.192	cu. in.

TIME.

60 seconds (*sec.*)	= 1 minute (*m.*).
60 minutes	= 1 hour (*hr.*).
24 hours	= 1 day (*da.*).
7 days	= 1 week (*wk.*).
30 days	= 1 month (*mo.*).
365 days	= 1 common year.
366 days	= 1 leap year.

CIRCULAR MEASURE.

Used for measuring angles and circumferences, reckoning latitudes and longitudes, etc.

60 seconds (″)	= 1 minute (′).
60 minutes	= 1 degree (°).
30 degrees	= 1 sign (*S.*).
12 signs or 360 degrees	= 1 circumference.

Apparently the sun makes an entire revolution of the earth * in 24 hours, and consequently travels 15° in 1 hour; hence,

1 hour of time	= 15° longitude.
1 minute of time	= 15′ longitude.
1 second of time	= 15″ longitude.

* *Really*, the revolution is that of the earth on its own axis.

MISCELLANEOUS TABLE.

12 units	= 1 dozen (*doz.*).
12 dozen	= 1 gross.
12 gross	= 1 great gross.
20 units	= 1 score.
24 sheets of paper	= 1 quire.
20 quires	= 1 ream.
196 lbs. flour	= 1 barrel.
200 lbs. pork	= 1 barrel.

When a sheet of paper is folded into two leaves, and a book made of this size, it is called a folio.

A sheet made into 4 leaves is called a quarto.

A sheet made into 8 leaves is called an octavo.

A sheet made into 12 leaves is called a duodecimo.